Bio-Based Cooling and Lubrication in Milling

By
Versace

TABLE OF CONTENTS

CHAPTER

1

Introduction

1.1 Background

Milling is a commonly used machining process for removing material from a workpiece surface. This technique utilizes a rotating cutter equipped with multiple cutting edges that interact with the workpiece to perform the material removal. Among its various forms, face milling is a specific method where the surface to be machined is oriented perpendicular to the axis of the rotating tool. Tool life plays a crucial role in this process, as it directly influences the quality of the surface finish and overall productivity. Enhanced tool longevity leads to better machining performance. Consequently, optimizing factors such as machining conditions, tool geometry, and operational settings is essential for extending tool life and achieving efficient outcomes. (Saleem and Mumtaz 2020; Reis et al. 2024). Nickel-based superalloys like Inconel 718 are difficult to machine because of their exceptional strength, low thermal conductivity, and the presence of carbides in their microstructure. These features lead to elevated cutting forces, rapid tool degradation, and reduced machinability. Traditional cooling techniques, such as flood cooling, have been widely used to counteract heat generation, but their adverse environmental effects and high operational costs have led to the exploration of alternative cooling strategies (Devillez et al. 2007; Mahesh et al. 2021).

Recent progress in cooling and lubrication methods, including Minimum Quantity Lubrication (MQL), have demonstrated effectiveness in machining hard-to-machine materials. Plant-based cutting fluids have gained attention as environmentally friendly substitutes for petroleum-derived coolants, owing to their biodegradability, reduced toxicity, and excellent lubricating properties. The vegetable oil-based coolant has been reported to reduce friction, lower tool wear, and improve surface quality, but its effectiveness compared to dry air cooling in machining Inconel 718 remains an open research question (Okafor and Nwoguh, 2020).

This research focuses on thoroughly assessing the machinability of Inconel 718 under dry conditions, compressed air, and vegetable oil-based lubricants by examining critical factors including tool wear, cutting force, thermal profile, and surface quality. The outcomes are expected to offer meaningful contributions toward eco-friendly machining methods and aid in formulating innovative cooling and lubrication techniques for processing superalloys.

1.2 Fundamental Aspects of Machining

Machining is a manufacturing process that shapes a workpiece to its final dimensions by eliminating excess material using a cutting tool. Material removal takes place due to plastic flow and shearing actions at the interface where the cutting tool engages with the workpiece. The area where this shearing and deformation takes place is referred to as the primary shear zone (Thimm et al. 2021). The excess material is expelled in the form of chips, which absorb most of the heat generated during machining. These chips slide along the tool's rake face, leading to an area where adhesion occurs, known as the secondary deformation zone. Additionally, there is another region where the tool's flank face makes contact with the machined surface, causing further deformation this is identified as the tertiary shear zone (Langenhorst et al. 2022).

Traditional machining processes include turning, boring, shaping, planning, drilling, milling, grinding, broaching, lapping, and honing. Cutting tools are important because they help remove material from the workpiece during machining. In the machining process, significant plastic deformation takes place in two main areas called the primary and secondary cutting zones, along with friction along the rake and flank faces of the tool (Zhang et al. 2021). The energy used for this deformation turns into heat. Additional thermal energy is generated both by the sliding action of the chip across the rake surface of the cutting tool and by the interaction between the tool and workpiece. To minimize heat buildup in both the primary and secondary cutting zones, cutting fluids are applied, serving both cooling and lubricating functions (Wang et al. 2023). The effectiveness of this process depends on the kind of cutting fluid used and how it is applied, which in turn affects the surface finish of the machined part.

Over the past few decades, sustainable manufacturing practices have gained significance, driven by productivity considerations and regulatory requirements from governmental bodies aimed at minimizing environmental impact. The demand for superalloys and hard-to-machine materials has surged, particularly in industries such as aerospace, nuclear power, marine engineering, and petrochemicals (Sinha et al. 2023). Consequently, various non-traditional machining techniques have emerged. Nevertheless, conventional machining methods continue to be widely utilized due to their efficiency, ability to achieve superior surface finishes, and capability to manufacture intricate components. In this regard, improving the efficiency of conventional machining greatly depends on selecting the appropriate tool material by considering factors such as the workpiece type, tool design, cutting conditions, and the chosen lubrication or cooling method.

Superalloys and specialized modern alloys pose machining challenges under standard conditions, necessitating the integration of advanced cooling technologies to mitigate temperature- related issues and improve productivity. Superalloys are advanced high-performance materials specifically designed to withstand extreme temperatures, both high and sub-zero, along with severe corrosion and fluctuating stresses.

1.3 Cutting Temperature in Machining

The substantial heat produced during metal cutting increases the temperature at the tool's cutting edge, which in turn directly influences how quickly the tool wears out. The majority of this heat is absorbed by the chip and carried away, while a smaller fraction is conducted into the workpiece. At the tool-chip interface, the region near the tool face experiences intense strain, causing continuous shearing of the workpiece material (Grzesik, 2020). As a result, the mechanical energy expended in this zone is entirely transformed into heat. Due to the small volume of material involved, extremely high temperatures are developed in this localized area as shown in Figure 1.1 (Korkmaz and Gupta 2024). On the rake face of the tool, the generated heat is steadily conducted away, leading to elevated temperatures that are influenced by factors such as the thermal conductivity of the workpiece material, tool geometry, machining conditions, and the type of cutting fluid used. As cutting speed increases, the temperature at the tool-chip interface also rises. At the point where the tool touches the workpiece, only a small amount of the total energy used turns into heat. Providing an adequate clearance angle on the cutting tool prevents direct rubbing of the newly machined surface against the tool, allowing most of the heat to be dissipated into the workpiece rather than being absorbed by the tool. This helps reduce thermal stress on the cutting tool.

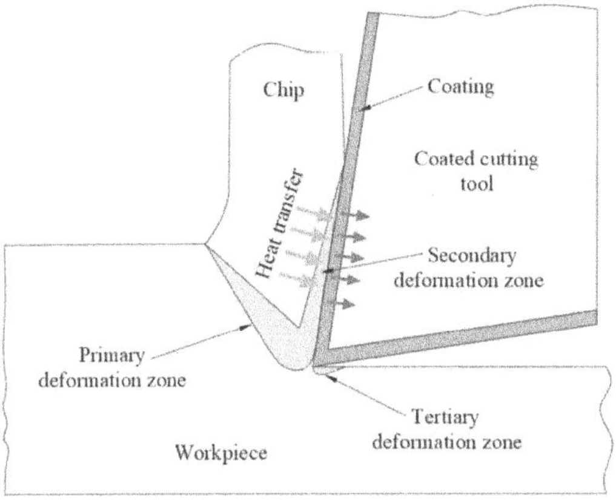

FIGURE 1.1 Cutting zone during machining, with permission of Korkmaz and Gupta., 2024 copyright Elsevier

1.4 Need for Metal Working Fluids

Traditional dry machining, though one of the oldest methods for metal cutting, is generally ineffective for hard materials like nickel-based superalloys due to excessive heat generation and rapid tool wear. The use of advanced cutting tool materials like cermet, ceramic, PCD, and CBN, along with modern coatings applied via PVD or CVD, has inhance tool life and surface finish under dry conditions. However, dry machining of superalloys still leads to high temperatures, reduced tool life, and increased power consumption, raising production costs (Palakudtewar and Gaikwad, 2014; Prasad et al. 2024).

To address this, cutting fluids are commonly used. Flood cooling helps lower cutting temperatures and reduce tool wear, extending tool life. At high cutting speeds, flood cooling becomes less effective as it cannot reach critical contact zones. Moreover, the use of synthetic and semi-synthetic coolants raises health concerns for operators and adds environmental and disposal costs, making sustainable alternatives increasingly important in modern machining. Minimum Quantity Lubrication (MQL) has emerged as an environmentally conscious alternative to conventional flood cooling, particularly suited for machining difficult-to-cut alloys. In MQL, a very small amount of lubricant typically in the range of 10–100 mL/hr is delivered in the form of fine aerosol droplets directly into the cutting zone using compressed air. The small amount of lubricant creates a thin film at the cutting zone that reduces friction and wear between the tool and workpiece (Dhar et al 2006).

The high-velocity air stream enables effective penetration of lubricant droplets into the tool–chip interface, improving chip evacuation and lowering localized temperature. This targeted and efficient mode of delivery not only enhances tool life and surface finish but also eliminates problems associated with coolant disposal, microbial growth, and operator exposure (Shokrani et al. 2012). Due to its low consumption, MQL offers economic advantages while aligning with sustainable manufacturing principles. These characteristics make MQL particularly attractive for machining nickel-based superalloys such as Inconel 718, where thermal management, lubrication efficiency, and environmental considerations are critical. A dedicated section on the MQL technique is covered in Chapter 2.

Cooling and lubrication conditions have a profound impact on the performance, efficiency, and sustainability of machining processes. Their primary roles include controlling cutting temperature, reducing friction at cutting zone, flushing away chips, and improving surface finish (Luo et al. 2024). In the absence of effective cooling or lubrication, excessive heat and friction are generated, which can lead to rapid tool wear, thermal damage to the workpiece, poor dimensional accuracy, and decreased tool life. The type and effectiveness of the cooling/lubrication method selected must therefore be aligned with the machining material, cutting parameters, and tool geometry to ensure optimal machining conditions. Figure 1.2 presents the various classifications of cutting fluids.

1.5 Current Scenario and Comparative Assessment of Biodegradable and Non-Biodegradable Cutting Fluids

Mineral-oil-based cutting fluids remain widely used in industry because of their low cost, good stability, and established supply chains. However, increasing concerns related to toxicity, poor biodegradability, and high disposal costs have driven a strong shift toward environmentally friendly alternatives. As a result, bio-based cutting fluids primarily vegetable oils (such as sunflower, canola, palm, castor, or jatropha) and engineered bio-esters are gaining significant attention as sustainable substitutes.

Biodegradable fluids generally offer superior lubricity due to their polar triglyceride structure, higher film strength, and lower toxicity, which often translate into reduced cutting forces, improved tool life, and better surface quality. Their main limitations include lower oxidative and thermal stability, higher viscosity, and reduced convective cooling capacity compared to mineral-oil emulsions. These drawbacks are commonly addressed through chemical modification (e.g., bio-ester synthesis), incorporation of additives (antioxidants or extreme-pressure agents), or dispersion of nanoparticles such as SiO_2, Al_2O_3, and graphene to create nano-enhanced biofluids with improved tribological and heat-transfer performance (Dennison et al., 2024).

Comparative studies consistently show that vegetable-oil-based fluids can reduce tool wear and cutting forces more effectively than mineral oils due to superior boundary lubrication. While mineral-oil emulsions still provide better bulk cooling, nano-enhanced biodegradable fluids substantially improve thermal conductivity and reduce cutting temperature. Surface finish is frequently equal or better with biodegradable fluids, especially when nanoparticle additives minimize built-up edge formation. Although mineral oils remain advantageous in terms of shelf life and initial cost, life-cycle and environmental considerations increasingly favour biodegradable alternatives (Afonso et al., 2022 Kazeem et al, 2022; Gan et al, 2024).

Emerging trends include the development of bio-esters with enhanced stability, advanced nanolubricants with superior thermal and frictional behaviour, and hybrid biodegradable formulations designed for machining difficult-to-cut materials such as nickel-based superalloys. Recent reviews and experimental studies report that properly formulated bio-based fluids can match or exceed the performance of conventional mineral-oil-based fluids across many machining metrics (Luo et al., 2024).

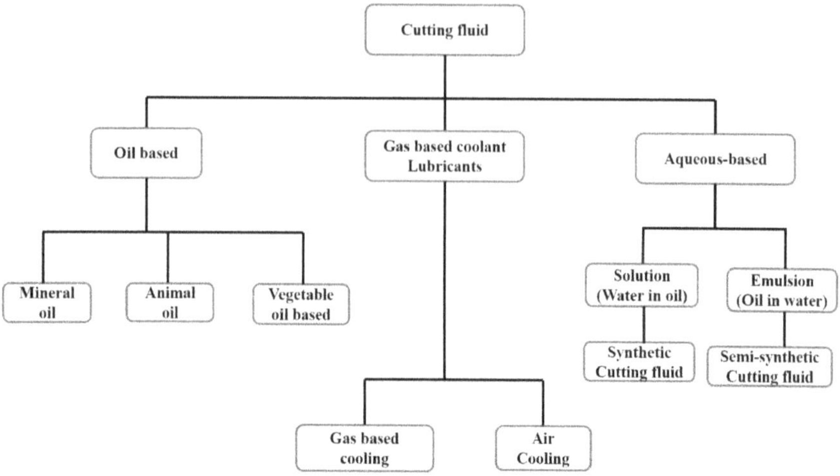

FIGURE 1.2 Types of cutting fluids

1.6 Putranjiva Roxburghii oil

Putranjiva roxburghii oil, extracted from the seeds of the Putranjiva tree, has garnered attention for its diverse applications and bioactive properties. The oil is rich in fatty acids such as oleic and linoleic acids, and contains bioactive compounds with antibacterial, antioxidant, anti-inflammatory, and potential anti-fertility effects, making it valuable in traditional medicine and as a bioactive reservoir for various health conditions (Emasushan and Britto, 2018). Recent studies have demonstrated its effectiveness in reversing fertility impairment in zebrafish models, where it improved ovarian follicle count, egg maturation, and sperm quality. Beyond medicinal uses, Putranjiva oil shows promise as a sustainable alternative to synthetic lubricants and cutting fluids, especially when modified with nanoparticles like silicon dioxide or graphene, which enhance its tribological and thermal properties for industrial applications.

1.7 Motivation and Research Aim

The growing need for advanced materials in sectors such as aerospace, automotive, nuclear, and power generation has driven the widespread adoption of nickel-based superalloys, particularly Inconel 718. It is widely used across high-temperature and high-performance engineering sectors due to its exceptional strength, corrosion resistance, and creep resistance up to about 700 °C. Specifically, Inconel 718 is employed in supersonic aircraft structures, gas turbines, rocket and aircraft engine components (including compressor blades), spacecraft parts, nuclear reactor components, and equipment for the chemical and oil and gas industries. In addition, the alloy's good weldability and mechanical stability at elevated temperatures make it suitable for manufacturing high-elasticity components such as springs and various critical parts used in aerospace and energy systems (Ravichander et al., 2022; Vereschaka et al, 2022).

Despite being renowned for its exceptional strength, ability to withstand high temperatures, and outstanding corrosion resistance, this alloy is considered difficult to machine. This is primarily due to its low ability to conduct heat, high tendency to harden during cutting, and strong interaction with tool materials. These factors contribute to excessive heat buildup, rapid tool degradation, and compromised surface quality, ultimately raising manufacturing expenses.

Conventional machining techniques like dry cutting and flood cooling come with notable limitations. Dry cutting leads to excessive heat accumulation, accelerating tool wear, while flood cooling, which typically utilizes synthetic or semi-synthetic oils, raises serious concerns regarding environmental safety and operator health. To meet the increasing demand for eco-conscious and economical manufacturing practices, alternative approaches such as Minimum Quantity Lubrication (MQL) using biodegradable vegetable oil-based lubricants have emerged as promising solutions. Despite this, there remains a lack of comprehensive studies comparing the effectiveness of dry and MQL machining of Inconel 718 under varying cooling methods. This study is driven by the objective to boost machining performance, extend tool service life, and achieve better surface quality, all while reducing ecological footprint. By evaluating the behavior of Inconel 718 during machining under both dry air and vegetable oil-assisted MQL conditions, this research aims to support the advancement of sustainable and high-efficiency machining technologies.

1.8 Organization of the Thesis

This thesis has been organized in seven chapters. The discussions of these chapters are briefly summarized as:

Chapter 1: Introduction

This chapter provides a comprehensive introduction of milling of nickel-based superalloys, specifically Inconel 718 and cutting fluids covering their characteristics, performance, and applications. It highlights the machining challenges associated with these 'difficult-to-cut' materials in comparison to conventional metals and alloys.

Chapter 2: Literature Review

This chapter presents an overview of previous studies focused on the formulation, characteristics, and uses of nanofluids derived from vegetable oils, with particular attention to Putranjiva roxburghii oil. It discusses the difficulties encountered during the machining of hard-to-machine materials such as Inconel alloys and emphasizes the need for environmentally friendly, high-efficiency lubrication solutions. The review compares conventional and nano cutting fluids in terms of tool wear, heat dissipation, and Surface integrity. Important elements such as mixture design, nanoparticle stability, lubrication efficiency, and thermal properties of nanofluids are examined, along with analytical techniques like FTIR and UV-Vis spectroscopy. The role of MQL and nano bio-lubricants in enhancing machining efficiency and reducing environmental impact is discussed. Additionally, studies using the G.E.P Box technique for optimizing machining parameters and tool geometry are included, aiming to improve sustainability and performance in precision machining.

Chapter 3: Materials and Experimentation Procedure

This chapter presents a comprehensive experimental framework, outlining the systematic approach adopted in the study. The experimental study is conducted in four separate stages, beginning with: (a) Production of a lubricant for machining operations using vegetable-oils and preparation of nanofluids by incorporating SiO_2 and Graphene-based nanoparticles. (b) Milling of Inconel 718 under dry conditions and an air-cooled environment to evaluate baseline machining characteristics. (c) Milling of Inconel 718 using a cutting fluid emulsion derived from Putranjiva roxburghii oil, applied through the MQL technique. (d) Milling of

Inconel 718 utilizing a nanoemulsion-based cooling approach to assess its effectiveness in improving machinability. A detailed explanation of the experimental procedure, data collection methodology, and data analysis techniques is provided in this chapter. Additionally, a comprehensive description of the machinery, equipment, and instruments employed in the study is documented to ensure clarity and reproducibility of the research.

Chapter 4: Characterizations of Putranjiva roxburghii Oil with Addition of SiO_2 and Graphene Nanoparticles

This chapter outlines the development, preparation, and detailed analysis of Putranjiva roxburghii-based nano cutting fluids. The fluids are prepared by dispersing nanoparticles into bio-based oil using surfactants and high-energy mixing to ensure stable dispersion. Tests are conducted to assess thermal, physical, and chemical properties, including thermal conductivity, density, viscosity, and rheological behavior. Wettability studies evaluate the spreading and adhesion properties of fluid on metal surfaces, influencing lubrication and cooling. FTIR identifies chemical bonds and structure, while UV-Vis spectroscopy examines optical properties and dispersion stability. These analyses highlight the potential of these eco-friendly nanofluids for efficient and sustainable machining, with improved lubrication, cooling, and tribological performance.

Chapter 5: Milling Performance of Inconel 718 under Dry, Air-Cooling, and Bio-oil based cooling with MQL Technique

This chapter investigates the milling performance of Inconel 718 under three distinct cooling strategies: dry machining, air-assisted cooling, and emulsion-based lubrication. Known for its exceptional mechanical strength, thermal stability, and resistance to corrosion, Inconel 718 poses significant machining challenges due to its rapid work-hardening behavior, poor heat dissipation, and requirement for elevated cutting forces. The study examines the influence of these cooling methods on critical machining parameters, including surface quality, thermal buildup, tool degradation, and overall process efficiency. A review of existing literature highlights the challenges involved in machining Inconel 718 and emphasizes the importance of effective cooling strategies. Dry machining is often associated with increased tool wear and thermal stress, while air cooling offers a more environmentally friendly approach with moderate improvements in heat dissipation. Among the approaches examined, the emulsion formulated with Putranjiva oil exhibited the most effective cooling and lubricating performance. Experimental results compare tool temperature, surface quality, and tool wear across all three conditions, providing valuable insights for optimizing the milling of Inconel 718 in industrial applications.

Chapter 6: Performance of Putranjiva Roxburghii Oil Based Nanoemulsions in Milling of Inconel 718

This chapter systematically evaluates the thermal, tribological, and machining performance of nano-doped bio-based cutting fluids during the milling of Inconel 718. The incorporation of SiO_2 and graphene nanoparticles into *Putranjiva roxburghii* oil-based emulsions led to significant improvements in key performance metrics, including thermal conductivity, specific heat, wettability, axial force, tool temperature, and surface finish. Nanofluids have recently gained attention for their superior cooling, lubrication, and heat transfer capabilities in contrast to conventional dry and wet approaches. This research explores how bio-derived and nanoparticle-enhanced cutting fluids influence tool degradation, surface quality, machining forces, heat generation, and chip morphology. The review of existing studies confirms the positive role of nanofluids and vegetable oil-based lubricants in boosting machining performance and promoting eco-friendly practices. The experimental setup includes milling operations with varied speeds, feed rates, and depths of cut, assessed through tool wear progression, surface quality, and axial force. Results show that nano-fluids outperform standard emulsions in reducing tool wear, improving surface finish, and enhancing thermal management. The chapter concludes by emphasizing the potential of MQL with bio-nanofluids for eco-friendly, high-performance machining of Inconel 718 and suggests further research on MQL optimization and hybrid cooling strategies.

Chapter 7: Conclusion and Future Scope

This chapter highlights the key contributions of the research and outlines potential future directions for further investigation in the field of machining optimization. The study has provided significant insights into the optimization of cutter geometry and process parameters for machining Inconel 718 under dry, air-cooled, and nanodoped bio cutting fluid environments using the G.E.P Box evolutionary technique. Through experimental procedures, this research has improved machining performance, reduced tool wear, and enhanced surface quality. The outcomes aid in promoting eco-friendly manufacturing techniques by demonstrating the effectiveness of nano-enhanced bio-based cutting fluids in reducing cutting forces and improving tool life. The results serve as a foundation for further research into the application of advanced optimization algorithms, hybrid cooling techniques, and real-time adaptive machining strategies. Further studies might focus on leveraging intelligent algorithms for live monitoring and control of operations, as well as assessing the effectiveness of dual-mode lubrication systems combining nanofluids with cryogenic cooling, and the development of environmentally friendly alternatives to conventional metalworking fluids. The implementation of digital twin technology and Industry 4.0 concepts in machining optimization can further enhance predictive accuracy and adaptive manufacturing capabilities. Overall, this research contributes to the evolution of precision machining by optimizing key machining parameters and promoting sustainable, high-performance manufacturing practices.

References

Afonso, I.S., Pereira, J., Ribeiro, A.E., Amaral, J.S., Rodrigues, N., Gomes, J.R., Lima, R. and Ribeiro, J., 2022. Analysis of a vegetable oil performance in a milling process by MQL lubrication. *Micromachines*, *13*(8), p.1254.

Barewar, S.D., Kotwani, A., Chougule, S.S. and Unune, D.R., 2021. Investigating a novel Ag/ZnO based hybrid nanofluid for sustainable machining of inconel 718 under nanofluid based minimum quantity lubrication. *Journal of Manufacturing Processes*, *66*, pp.313-324.

Dennison, M.S., Jebabalan, S.K. and Barik, D., 2024. Applicability of nano-cutting fluids for enhanced cooling, low tool wear, and high tribological performance during machining—a review. *Discover Applied Sciences*, *6*(12), p.663.

Devillez A, Schneider F, Dominiak S, Dudzinski D and Larrouquere D 2007 Cutting forces and wear in dry machining of Inconel 718 with coated carbide tools. Wear 262(7-8): 931-942.

Dhar, N. R., Paul, S., & Chattopadhyay, A. B. (2006). Machining of AISI 4140 steel under minimum quantity lubrication (MQL) and conventional cooling: A comparative study. Journal of Materials Processing Technology, 172(2), 299-304.

Dos Reis, A., Fernandes, G.H.N., de Sousa, J.A.G., Vaughan, L.L.T., Cangue, F., Machado, A.R. and Guesser, W.L., 2024. Tool life assessment of high strength cast iron alloys in dry face milling operations. Journal of Manufacturing Processes, 111, pp.180-193.

Emasushan, M. and Britto, J.S., 2018. Seed oil extraction and phytochemical analysis of Putranjiva roxburghiiwall. (Putranjivaceae). *Journal of Emerging Technologies and Innovative Research*, *5*(10).

Gan, C.K., Liew, P.J., Leong, K.Y. and Yan, J., 2024. Biodegradable cutting fluids for sustainable manufacturing: a review of machining mechanisms and performance. *The International Journal of Advanced Manufacturing Technology*, *131*(3), pp.955-975.

Grzesik, W., 2020. Modelling of heat generation and transfer in metal cutting: a short review. *Journal of Machine Engineering, 20.*

Ishfaq, K. and Rehman, M., 2023. Machinability investigation using Cu-mixed biodegradable dielectric for eco-friendly/sustainable machining of Ni-based superalloy. *The International Journal of Advanced Manufacturing Technology*, *125*(5), pp.2243-2264.

Kazeem, R.A., Fadare, D.A., Ikumapayi, O.M., Adediran, A.A., Aliyu, S.J., Akinlabi, S.A., Jen, T.C. and Akinlabi, E.T., 2022. Advances in the application of vegetable-oil-based cutting fluids to sustainable machining operations—a review. *Lubricants*, *10*(4), p.69.

Korkmaz, M.E. and Gupta, M.K., 2024. Nano lubricants in machining and tribology applications: A state of the art review on challenges and future trend. *Journal of Molecular Liquids*, *407*, p.125261.

Langenhorst, L., Sölter, J. and Heinzel, C., 2022. Partitioning of primary shear zone heat in face milling. *CIRP Annals*, *71*(1), pp.53-56.

Luo, X., Wu, S., Wang, D., Yun, Y., An, Q. and Li, C., 2024. Sustainable development of cutting fluids: the comprehensive review of vegetable oil. *Journal of Cleaner Production*, *473*, p.143544.

Mahesh, K., Philip, J.T., Joshi, S.N. and Kuriachen, B., 2021. Machinability of Inconel 718: A critical review on the impact of cutting temperatures. *Materials and Manufacturing Processes*, *36*(7), pp.753-791.

Okafor, A.C. and Nwoguh, T.O., 2020. Comparative evaluation of soybean oil–based MQL flow rates and emulsion flood cooling strategy in high-speed face milling of Inconel 718. *The International Journal of Advanced Manufacturing Technology*, *107*(9), pp.3779-3793.

Palakudtewar, R.K. and Gaikwad, S.V., 2014. Dry machining of superalloys: difficulties and remedies. *Int. J. Sci. Res*, *3*(7), pp.277-282.

Prasad, G., Vijay, G.S., Kamath, R.C. and Hemmady, H.J., 2024. Optimization of the tool wear and surface roughness in the high-speed dry turning of Inconel 800. *Cogent Engineering*, *11*(1), p.2308993.

Ravichander, B.B., Rahimzadeh, A., Farhang, B., Shayesteh Moghaddam, N., Amerinatanzi, A. and Mehrpouya, M., 2021. A prediction model for additive manufacturing of Inconel 718 superalloy. *Applied Sciences*, *11*(17), p.8010.

Saleem, M.Q. and Mumtaz, S., 2020. Face milling of Inconel 625 via wiper inserts: Evaluation of tool life and workpiece surface integrity. *Journal of Manufacturing Processes*, *56*, pp.322-336.

Shokrani, A., Dhokia, V., & Newman, S. T. (2012). Environmentally conscious machining of difficult-to-machine materials with regard to cutting fluids. International Journal of Machine Tools and Manufacture, 57, 83-101.

Sinha, M.K., Pal, A., Kishore, K., Singh, A., Archana, Sansanwal, H. and Sharma, P., 2023. Applications of sustainable techniques in machinability improvement of superalloys: a comprehensive review. *International Journal on Interactive Design and Manufacturing (IJIDeM)*, *17*(2), pp.473-498.

Sinha, M.K., Pal, A., Kishore, K., Singh, A., Archana, Sansanwal, H. and Sharma, P., 2023. Applications of sustainable techniques in machinability improvement of superalloys: a comprehensive review. *International Journal on Interactive Design and Manufacturing (IJIDeM)*, *17*(2), pp.473-498.

Thimm, B., Glavas, A., Reuber, M. and Christ, H.J., 2021. Determination of chip speed and shear strain rate in primary shear zone using digital image correlation (DIC) in linear-orthogonal cutting experiments. *Journal of Materials Processing Technology*, *289*, p.116957.

Vereschaka, A., Milovich, F., Andreev, N., Migranov, M., Alexandrov, I., Muranov, A., Mikhailov, M. and Tatarkanov, A., 2022. Specific Application Features of Ti-TiN-(Ti, Cr, Al) N, Zr-ZrN-(Zr, Mo, Al) N, and ZrHf-(Zr, Hf) N-(Zr, Hf, Cr, Mo, Al) N Multilayered Nanocomposite Coatings in End Milling of the Inconel 718 Nickel-Chromium Alloy. *Journal of Composites Science*, *6*(12), p.382.

Wang, B., Liu, Z., Cai, Y., Song, Q. and Ren, X., 2023. Effects of tool angles and uncut chip thickness on consumption of plastic deformation energy during machining process. *Journal of Manufacturing Processes*, *87*, pp.123-132.

Yan, P., Rong, Y. and Wang, G., 2016. The effect of cutting fluids applied in metal cutting process. *Proceedings of the Institution of Mechanical Engineers, Part B: Journal of Engineering Manufacture*, *230*(1), pp.19-37.

Zhang, D., Zhang, X.M., Nie, G.C., Yang, Z.Y. and Ding, H., 2021. Characterization of material strain and thermal softening effects in the cutting process. *International Journal of Machine Tools and Manufacture*, *160*, p.103672.

CHAPTER

2

Literature review

2.1 Introduction

Milling is a widely used and essential machining operation that involves a rotating multi-edge cutter removing material from a fixed or moving workpiece to create parts with specific shapes and surface finishes. It is particularly suited for generating flat surfaces, slots, contoured profiles, and complex three-dimensional features in a wide range of materials. The milling process's efficiency and quality are determined by several crucial machining parameters, including the cutting speed, depth of cut, and feed rate. The speed at which the milling tool rotates known as cutting speed is a key factor in heat buildup and significantly affects the durability of the cutting tool. Meanwhile, the feed rate, representing how quickly the workpiece moves in relation to the cutter, has a major influence on the quality of the surface finish, the shape of the chips produced, and the magnitude of the cutting forces. The thickness of material removed in a single pass, known as the depth of cut, has a direct effect on both the speed of material removal and the mechanical stress placed on the cutting tool. Alongside this, factors such as the tool's design specifically its rake angle, clearance angle, and nose radius the chosen cooling and lubrication method, and the overall stiffness of the machine-tool-workpiece assembly play vital roles in determining machining effectiveness.

These elements together influence essential outcomes like surface quality, precision in dimensions, tool degradation, cutting force, vibrations, and the overall effectiveness of the machining process.

Machining becomes particularly challenging when dealing with superalloys primarily based on cobalt, nickel, or nickel-iron which are engineered to retain high mechanical strength, oxidation resistance, and structural integrity under extreme thermal and mechanical loads. Their application is widespread in industries demanding superior heat and corrosion resistance, such as aerospace gas turbines, coal-fired power plants, nuclear reactors, and chemical processing units (Seenath and Sarhan, 2024). These alloys exhibit low machinability owing to their exceptional strength, tendency to harden during machining, and limited ability to conduct heat. Consequently, considerable efforts have been made over the years to enhance the machining performance of nickel-based superalloys such as Inconel 718 by refining cutting conditions, selecting suitable tool materials and coatings, and implementing advanced lubrication and cooling strategies. These efforts aim to reduce tool wear, improve surface integrity, and enhance productivity while also addressing environmental and economic concerns.

This chapter offers an in-depth overview of previous studies related to the milling of nickel-based superalloys, emphasizing Inconel 718. Renowned for its outstanding mechanical strength and thermal resistance, Inconel 718 is extensively applied in the aerospace and energy industries, yet presents notable difficulties during machining. The discussion highlights the influence of different cooling methods on key milling aspects, such as surface quality, tool degradation, heat buildup, and the behavior of cutting forces. Section 2.2 reviews existing literature on the machinability of nickel-based alloys. Section 2.3 examines how various cooling approaches, including dry cutting and air-assisted cooling, affect the machining behavior of Inconel 718, highlighting the eco-friendly nature of compressed air in heat and chip removal. Section 2.4 focuses on the role of various cutting fluids and nanofluids, emphasizing their application and effectiveness in machining operations. Section 2.5 discusses the performance of machining processes in relation to various process parameters. Section 2.6 presents a summary of the reviewed literature, while Section 2.7 identifies major research gaps that remain unaddressed. Finally, Section 2.8 outlines the objectives of the present thesis, formulated based on the insights gained from the literature review. The chapter compares different cooling approaches, outlining their pros and cons, and provides insights into selecting optimal, environmentally responsible techniques for milling Inconel 718 efficiently.

2.2 Machinability Studies on Nickel-based Alloys

Machining is a vital process within the manufacturing sector, crucial for shaping components to the desired dimensions, geometries, and tolerances required for diverse applications, including those in the aerospace and defence sectors. The materials employed in these fields often include titanium and nickel-based superalloys, known not only for their exceptional mechanical and thermal properties but also for their machining challenges. Examples of such difficult-to-machine materials are Inconel 625, Nimonic 90, Nimonic 80A, Inconel 718 and Ti6Al4V. Nickel-based superalloys, in particular, exhibit outstanding creep resistance under high-temperature conditions (Zhang and Xu, 2021a). These very qualities, however, contribute to their reputation as some of the most difficult materials to machine. Despite advancements in machining methods, cutting tools, and inspection systems, the machining of these alloys remains a complex challenge due to the high stresses generated during processing and the stringent requirements for finished part quality. To achieve the desired balance between product quality, efficiency, and cost-effectiveness, the careful selection of machining parameters is essential. Extensive research has been carried out to explore various aspects of the machinability of nickel-based superalloys, particularly in identifying optimal cutting conditions. Recent investigations have built upon earlier findings, offering new insights and reevaluating prior conclusions in response to ongoing concerns in machining these materials (Kaya and Akyüz, 2017).

Nimonic 90, an alloy composed of nickel, chromium, and cobalt, is extensively utilized in turbine components across various applications, including aircraft, submarines, and steam as well as nuclear power systems. Its superior resistance to corrosion, oxidation, and creep up to 920 °C, along with its high strength at elevated temperatures, contributes to its poor machinability. These properties often lead to substantial wear of the cutting tools, negatively impacting productivity and increasing manufacturing costs (Behera et al., 2016). In addition, the poor heat conduction properties of these superalloys lead to intense heat buildup and increased friction at the tool–workpiece interface, which speeds up tool wear and may compromise the surface quality of the machined part. Inconel 625, a nickel-based superalloy, is well-known for retaining its strength and toughness over a wide temperature spectrum from cryogenic levels up to 1000 °C. Its composition includes elements that resist corrosion, oxidation, and magnetism, making it exceptionally resistant to fatigue and stress corrosion cracking. These characteristics make it a preferred choice in demanding environments such as gas turbines, aerospace systems, chemical processing units, nuclear reactors, high-temperature furnaces, and offshore structures. Similar to Inconel 718, the optimal milling speed for Inconel 625 ranges between 6 and 9 meters per minute, with a maximum feed rate of 0.15 mm per tooth (Ezugwu et al., 1999; Ezugau et al., 2005).

Thakur and Gangopadhyay (2016) carried out a comprehensive review centered on surface integrity during the machining of different nickel-based superalloys. Results demonstrated that both the rate of cutting and the level of tool engagement contribute to changes in the surface texture, their effects are further impacted by conditions like tool wear and the development of BUE. Among all parameters, feed rate emerged as the most critical factor influencing surface quality. The study also highlighted the crucial role of proper lubrication at the area where the machining tool meets the component in achieving efficient machining performance across various nickel alloys.

2.2.1 Machining of Inconel 718 alloy

Inconel 718 is being used more and more because it can handle very high temperatures and has strong mechanical and chemical properties. That is why it is often used in airplanes, nuclear plants, and ships. However, during its machining, a lot of heat is produced where the tool touches the metal. This heat causes the tool to wear out faster and can damage the surface of the part. Table 1 presents the detailed chemical composition of Inconel 718 alloy (Memarianpour et al., 2021).

Table 2.1 Chemical composition of Inconel 718 alloy

Element	Compositions (Wt.%)	Element	Compositions (Wt.%)
Ni	53.4	Si	0.12
Cr	18.8	Mn	0.07
Cb	5.27	Cu	0.07
Mo	2.99	C	0.03
Ti	1.02	P	0.01
Al	0.5	Fe	Balance
Co	0.17		

Pusavec et al. (2014) utilized Response Surface Methodology (RSM) and Analysis of Variance (ANOVA) to model and optimize the machining performance of Inconel 718 during turning with carbide cutting tools. Their study identified feed rate as the primary factor influencing surface roughness (R_a), while both depth of cut and feed rate played significant roles in determining cutting force. Similarly, Farshid et al. (2014) focused on optimizing machining conditions to reduce tensile residual stresses during the turning of Inconel 718. Ezugwu et al. (2005) explored the machining behavior of coated cemented carbide tools in high-speed turning operations involving Inconel 718, while Arunachalam et al. (2004) explored how cutting tool geometry affected surface quality during face turning. Their findings revealed that increasing the nose radius enhanced surface finish, and the application of coolant reduced surface roughness. Fan et al. (2013) investigated the surface finish during turning with coated carbide tools and found that the development and subsequent removal of built-up edges played a key role in determining surface quality.

Fan et al. (2015) investigated the behaviour at the interaction occurring at the interface of the cutting tool and the generated chip during the turning process. Inconel 718 was machined by TiAlN tool. Tool life got enhanced under the moist air condition as compared to dry cutting condition. The oxides Cr_2O_3, Fe_2O_3, and Fe_3O_4 were generated during lubrication that prohibited the tool wear. Sarkar and Datta (2021) investigated the turning performance of Inconel 718 under dry, minimum quantity lubrication (MQL) and nanofluid MQL (NFMQL). The flank wear and tool tip temperature were examined. Beyond 83 m/min cutting speed, the flank wear was higher in NFMQL than that in MQL condition. The tool tip temperature could be reduced by 60% in MQL and 67% in NFMQL, as compared to dry condition.

Saleem and Mehmood (2022) experimentally investigated the surface integrity and tool behaviour in dry and MQL turning of Inconel 718 with sunflower and castor oils as the cutting fluids. The most desirable centreline average (CLA) surface roughness (0.43 μm) and minimum tool wear (116 μm) were obtained using sunflower oil as a cutting fluid in MQL condition. These were 52% and 26% lower, respectively, as related to dry machining under the same conditions. Ross et al. (2022) studied the surface morphological properties in the milling of Nimonic 80A alloy in a hybrid (CO_2 + MQL) cooling technique that offered sufficient cooling and lubrication. The hybrid setup reduced the temperature and specific cutting energy by 34–53% and 17–19%, respectively, as compared to the MQL setup.

Manimaran et al. (2021) examined how the combined cryogenic and minimum quantity lubrication (CMQL) method influenced surface finish, texture, residual stress patterns, microhardness, and tool degradation. A key advantage identified with CMQL, compared to conventional techniques, was its ability to produce surfaces exhibiting compressive residual stress characteristics. An increase of 4–9% in microhardness of the surface and a 22–71% reduction in tool wear were reported. Periane et al. (2022) carried out a tool life test for milling of Inconel 718 under dry and MQL machining conditions. Tool life assessments were performed using Inconel 718 specimens fabricated through both cast and wrought (C&W) and SLM (selective laser melting) traditional casting. The study also investigated tool wear progression. In both processes, flank wear and cutting-edge fractures were the primary failure modes for reducing tool life. Results indicate that the microstructural differences C&W and SLM selective laser melting samples had a greater impact on tool life than the machining environment. While machining SLM samples, tool life was 80% longer in dry and 43% longer in MQL compared to C&W samples.

Pinheiro et al. (2021) analyzed how different machining parameters influenced surface finish, energy usage, tool wear, and vibration during the turning of Inconel 718, employing two CNMG120408 carbide inserts with PVD coatings of TiAlN/TiN. Employing an RSM-based desirability function, the most favorable machining parameters determined under MQL were a feed rate of 0.1 mm/rev and a cutting velocity of 69.9 m/min. Jadam et al. (2021) investigated the machinability of Inconel 718 by evaluating various cutting tool materials, including uncoated carbide, PVD-coated cermet inserts with TiN/TiCN/TiN layers, and PVD TiAlN-coated PCBN carbide tools. Their findings showed that cermet inserts delivered reduced cutting forces and improved surface quality.

2.2.2 Machining of other Nickel based alloys

Nickel-based alloys like Hastelloy, Monel, Incoloy, and Waspaloy exhibit machining difficulties comparable to those of Inconel 718, primarily due to their good strength, toughness, and low heat conductivity. Commonly utilized in sectors such as aerospace, marine, and chemical processing for their outstanding resistance to corrosion and high temperatures, these materials often lead to increase tool wear, cutting forces, and significant heat generation during machining operations.To overcome these issues, optimized cutting parameters, advanced tool materials (like coated carbide or ceramics), and efficient cooling or lubrication strategies, such as MQL or cryogenic cooling, are essential for improving machinability and surface integrity.

2.2.3 Surface Integrity analysis of machining of Incoel 718

The surface integrity is strongly influenced by tool material, machining parameters, and tool wear. The studies reported that new PCBN and carbide tools are capable of generating comparable and satisfactory surface finish; however, tool wear significantly degrades surface quality when using coated carbide or ceramic tools, leading to increased roughness, wavy profiles, and typical defects such as smearing, laps, tearing, and micro-cracks. In contrast, PCBN tools maintain more stable surface topography even in worn conditions. Subsurface evaluations reveal the formation of a severely plastically deformed ultrafine-grained layer, followed by elongated grains and slip bands, with tool wear intensifying deformation and altering grain size. Near-surface hardness increases noticeably due to severe strain hardening, with worn tools producing hardness values significantly higher than the bulk material. Residual stress analysis further indicates that machining induces substantial tensile stresses, which escalate sharply with tool wear particularly for PCBN and ceramic tools at high cutting speeds while carbide tools generally produce more favorable stress states. Together, these findings have been integrated into the thesis to more clearly demonstrate how machining conditions affect surface roughness, microstructure, hardness, and residual stresses in Inconel 718 (Saoubi et al. 2012; Akhtar et al. 2016).

Ezhilarasan et al. (2013) evaluated the turning behavior of Nimonic-263 alloy employing whisker-reinforced ceramic inserts in a dry machining. The study concluded that feed rate significantly impacts both the superiority of the machined surface and the magnitude of cutting forces, while tool wear is affected by a combination of feed rate and depth of cut. In a separate study, Davoodi and Eskandari (2015) examined the impact of machining variables on the metal removal efficiency and lifespan of coated carbide inserts while processing N-155 alloy. The results demonstrated that both the individual and interactive effects of cutting speed and feed rate were key factors affecting tool life. Junyun & Qingliang (2015) have formulated a predictive model for surface roughness when machining other work materials using a diamond cutting tool. Similarly, Motorcu et al. (2014) analyzed how variations in spindle speed, material feed, and different tool types affected the quality of surface texture while drilling Waspaloy. Meanwhile, investigations on machining the C-263 superalloy have shown that cutting forces escalate significantly as the feed rate increases (Ezilarasan et al., 2014; Zhang et al. 2015). Nath et al. (2015) conducted machinability experiments on nickel alloys during face milling using coated carbide tools. Their study optimized machining parameters based on material removal rate (MRR) and cost of machining.

Gomez-Parra et al. (2013) established that the formation of a BUE has a direct impact on surface morphology. Sharman et al. (2015) examined how tool nose radius affects surface integrity and residual stress in turning Inconel 718. Larger radii (2–6 mm) increase plastic deformation and surface tensile stress (up to 1550 MPa), especially with worn tools, while also deepening compressive stress below the surface. Though larger radii improve productivity, they raise thermal stress due to chip thinning, risking component integrity. Hence, they are not ideal for finishing operations despite better cutting efficiency. Zahrani and Sedghi (2014) investigated the turning of Monel K-500 using a coated carbide tool, focusing on surface quality, tool wear, and dimensional accuracy. The results showed that feed rate and nose radius had a statistically significant impact on surface finish (P-value < 0.05). Antonialli et al. (2020) examined the machining characteristics of the Ni-30 iron-based superalloy designed for high-temperature applications. Their research concluded that the best outcomes were achieved with high-pressure coolant application, tools featuring a negative rake angle, and higher cutting speeds. Additionally, the dominant forms of tool wear were identified as abrasion and attrition, indicating that the use of hard-coated tools could improve machining performance.

Singh et al. (2021) experimentally studied the machining performance of Hastelloy C-276 alloy under three lubrication environments: dry cutting, flood cooling condition, and MQL using vegetable oil. The study found that dry machining produced highly serrated chips, while the use of vegetable oil-based MQL considerably lowered cutting temperatures, improved surface finish, and decreased the chip reduction coefficient (CRC). The enhancements resulted from superior thermal regulation and lubricating action by the air-assisted delivery of vegetable oil directly into the cutting area. Gupta et al. (2021) explored the cylindrical turning of Inconel 800 alloy under dry, MQL, NMQL, and nitrogen-assisted cooling using a CNMG 120408 carbide insert. They determined that total energy consumption was reduced by 11.3%, 9.4%, and 7.5% under nitrogen cooling, NMQL, and MQL, respectively, compared to dry machining. Additionally, tool wear was lowered by 46.6%, 36.3%, and 12.1% with nitrogen cooling, NMQL, and MQL, respectively. Singh et al. (2022) conducted a review on eco-friendly machining approaches for nickel-based alloys, evaluating the performance differences between coated and uncoated cutting tools. Their findings emphasized the necessity of lubrication and cooling in machining nickel-based superalloys, particularly under demanding conditions, and highlighted that coatings generally enhance machining efficiency. MQL and cryogenic cooling were found to provide superior performance due to their enhanced lubrication and cooling characteristics.

2.3 Cooling and Lubrication Strategies in Machining: A Comparative Overview

The selection of an appropriate cooling strategy during machining has a deep influence on tool wear, surface integrity, energy consumption, and environmental sustainability. Numerous studies have compared traditional and advanced cooling approaches, emphasizing the superiority of modern techniques over conventional dry and flood cooling systems. Cryogenic cooling typically utilizing liquid nitrogen or carbon dioxide has garnered considerable attention for its capability to minimize tool degradation and enhance surface quality, energy consumption, and machining costs, while simultaneously minimizing carbon emissions (Zlatanovic et al., 2022; Gupta et al., 2014). This method is especially advantageous for difficult-to-machine materials such as Inconel alloys, hardened steels, and elastomers, as it enhances machinability by modifying material properties at cryogenic temperatures (Rakesh and Chakradhar 2023).

The performance of machining operations on nickel-based superalloys like Inconel 718 is greatly affected by the cooling or lubrication method used. Due to its poor ability to conduct heat and its inclination to harden during cutting, Inconel 718 accumulates substantial heat in the cutting area. Hen Zhang et al. 2015ce, choosing an effective cooling strategy is essential for enhancing the machinability of this material. Minimum quantity lubrication (MQL), particularly when integrated with nanoparticles (i.e., nanofluid MQL), has also been extensively studied as a sustainable alternative to flood cooling. Researchers have reported that MQL and nanofluid MQL significantly improve tool life, reduce cutting temperatures, and enhance surface finish, while consuming minimal lubricant and reducing environmental impact (Pimenov et al. 2024). Another innovative approach involves internally cooled cutting tools, which offer localized heat removal without the need for external fluids, thus contributing to cleaner machining environments. Depending on the material being machined and selected process settings, such systems can occasionally cause higher cutting loads (Fernandes et al., 2023).

Although significant progress has been made, studies show that there is no one-size-fits-all cooling approach. The performance of any cooling technique largely varies based on the composition of the work material, specific cutting parameters, and environmental sustainability goals. Nevertheless, sustainable cooling methods such as cryogenic cooling and MQL are increasingly favored for their balanced performance in terms of machining efficiency and ecological impact (Pimenov et al.2024).

2.3.1 Machining in dry conditions

Dry machining refers to a process where cutting fluids and their delivery systems are completely excluded, in contrast to methods like MQL, wet cooling, or cryogenic techniques. Without lubricants, the tool and workpiece heat up more during machining (Mia et al., 2018; Pimenov et al., 2022b). This rise in temperature is mainly due to increased friction and limited heat dissipation, as air is a poor conductor. Effective implementation of dry machining depends on choosing appropriate tool materials that can withstand wear. Tools suited for this process must possess high hardness, resistance to extreme pressure and temperature, chemical stability, and strong thermal fatigue endurance (Şirin et al., 2021). This approach offers several benefits, as highlighted by Salman et al. (2018). It eliminates the need for additional equipment such as high-pressure pumps, storage tanks, and mixing chambers for cutting fluid application. Moreover, there is no requirement for systems related to the collection, filtration, recycling, or disposal of cutting fluids, thereby simplifying the overall process. Chip handling and disposal become more efficient, and the absence of cutting fluids contributes to a cleaner workplace, reduced environmental impact, and minimized health risks for operators. Additionally, equipment maintenance is simplified, water consumption is significantly lowered, and overall machining costs are effectively reduced.

Although dry machining offers certain benefits, its application to nickel-based superalloys leads to substantial heat accumulation at the cutting interface, which in turn causes rapid tool deterioration, the formation of BUE, and a decline in surface quality. Domenico et al. (2013) conducted research on dry machining of Inconel 718 using coated carbide tools and found that surface quality could be enhanced by employing cutting speeds above 70 m/min in combination with a lower feed rate. Devillez et al. (2011) explored the performance of CVD TiCN-Al_2O_3-TiN coated carbide tools during dry machining of Inconel 718. The investigation indicated that 60 m/min cutting velocity led to optimal machining conditions, a satisfactory surface finish could be achieved with microhardness levels in the machining-affected zones comparable to those observed in wet machining. They also observed that under dry machining environment, surface finish was affected by the presence of residual built-up edges, which formed because of the significant rise in temperature in the cutting zone. In their detailed review on sustainable practices in metal cutting, Pervaiz et al. (2018) discussed the difficulties encountered during dry machining of materials such as nickel and titanium alloys, which are known for being hard to machine. They emphasized that excessive heat generation during machining led to poor dimensional accuracy, reduced tool life, and inferior surface finish. Additionally, they pointed out that the production of metallic dust in dry machining could pose health hazards to operators.

2.3.2 Effect of compressed air/cryogenic cooling

A sustainable alternative approach has been identified to manage heat and lubrication at the machining interface, which helps reduce the need for traditional coolants and their harmful effects (Liew et al., 2017). This method is called the air or gas cooling technique. Instead of using liquid coolants, it uses gases such as normal air, nitrogen (N_2), oxygen (O_2), carbon dioxide (CO_2), argon, and even water vapour to reduce heat and friction during cutting. One of the biggest advantages of this technique is that it is low-cost, especially because air is naturally available in large amounts and does not need to be specially prepared or handled like chemical coolants.

Compressed air is very useful in situations where liquid coolants cannot easily reach the cutting area, such as in deep or complex machining operations. It helps clear away chips and cool the tool effectively (Shokrani et al., 2012). In some cases, chilled air, which is cooled to low temperatures before being applied can also be used. This type of cooling helps to increase the life of the cutting tool without causing any rise in the cutting forces (Debnath et al., 2014). This means that the tool stays sharp for longer period and the machining process remains stable, while eliminating the ecological and occupational risks typically associated with standard lubricants.

2.3.3 Effect of cutting fluids on machining

The industrial transformation took root in the mid-1700s, marking a significant shift driven by the application of scientific advancements to manufacturing practices. During this period, steel products became an essential part of everyday life, leading to a substantial rise in their demand (Boxall, 2019). Producing steel components required extensive use of metalworking machinery. These machines employ cutting tools to remove excess material and achieve the desired geometry. However, this process often results in dimensional deviations and unsatisfactory surface quality. According to various studies, tool wear is identified as the primary reason behind these issues. Several elements contribute to tool deformation, including mechanical stress, elevated temperatures during cutting, excessive depth of cut and feed rates, intense friction at the tool–chip contact area, and the formation of BUE (Shokoohi, 2019; Gajrani and Sarkar 2020a).

Mahesh et al. (2021) highlighted the challenges from high cutting temperatures that cause tool wear and surface damage. They emphasized the significance of environmentally sustainable approaches such as dry cutting and high-pressure coolant application. They recognized that textured tools help minimize heat generation and friction, providing effective means to enhance tool longevity and surface finish. The high temperatures accelerate tool degradation and negatively affect the surface integrity, ultimately diminishing overall machining efficiency. The study by Ogedengbe et al. (2019) found that machining-induced heat altered the texture of the machined surface. Elevated temperatures were found to shorten tool life and lead to a deterioration in surface quality. The thermal load generated in cutting operations is a key factor influencing the surface profile of the final product. This characteristic is particularly important in component design, as it impacts critical performance metrics like resistance to wear and fatigue durability.

The excessive heat produced during machining can lead to further challenges such as thermal distortion, which affect the dimensional accuracy of the component as well as surface degradation through oxidation, burning, and accelerated corrosion. Fundamentally, the cutting operation-induced heat production during machning operations diminishes tool lifespan and adversely affects the workpiece's structural and surface properties, which can ultimately result in early failure of the manufactured components (Khajehzadeh and Razfar, 2020). When processing soft and ductile metals, significant friction between the chip and the cutting tool often causes small material fragments to build up on the rake surface near the cutting edge. Over time, this accumulation forms a mass that acts as a false cutting edge, a phenomenon referred to as BUE. BUE formation is commonly seen during operations involving single-point tools and can lead to numerous problems such as reduced tool longevity, inferior surface finish, dimensional variations, and other machining inconsistencies. The occurrence of BUE is often linked to low cutting speeds or improper feed rates (Song et al., 2020).

During machining, most of the generated heat is removed through the chips, while the rest is absorbed by the cutting tool. Managing temperature effectively and ensuring adequate lubrication are essential to prevent thermal damage and extend the tool's life. This is typically achieved by applying cutting fluids or metalworking fluids (MWFs) directly to the cutting zone. These fluids play a pivotal role in optimizing the metal cutting process. Initially, their primary function was to provide cooling during machining operations. As the fluids enter the cutting area, they help dissipate heat and facilitate chip evacuation. Cutting fluids serve multiple functions, including cooling, lubrication, cleaning of the machining area, and corrosion prevention (Sen et al., 2021; Yang et al., 2021). Essential characteristics like surface tension and viscosity play a vital role in removing chips and debris, thereby minimizing the need for manual intervention (Kantharaj et al., 2021). Cutting fluids also enhance lubrication between the tool and workpiece, allowing for higher cutting speeds. This improved lubrication helps lower tool wear and leads to a smoother machined surface and dimensional precision of the workpiece, leading to higher productivity and fewer defective parts compared to dry machining (Debnath et al., 2014; Bayat et al., 2021).

A fine layer of cutting fluid develops along the cutting edge of tool and the workpiece surface, effectively reducing heat and enhancing lubrication. This is especially advantageous during low-speed machining, where friction is more pronounced. At higher speeds, cutting fluids assist in suppressing the development of BUE on the tool. According to their chemical composition, cutting fluids are typically divided into two main types: oil-based and water-based (Zheng et al., 2023).

2.4 Cutting Fluids: Classifications and Cooling Techniques

During material processing, the application of cutting fluids plays an essential role in dissipating the heat generated during machining, thereby lowering the cutting temperature. Additionally, these fluids help minimize friction and tool wear, ultimately enhancing the surface integrity of the workpiece. Cutting fluids are thus considered an essential element in the manufacturing sector. Despite their benefits, conventional cutting fluids derived from mineral oils often contain harmful substances that pose risks to human health and contribute to environmental pollution. Consequently, there is growing global interest among researchers in developing sustainable, eco-conscious substitutes. In this context, vegetable oils have emerged as a strong alternative due to their biodegradability and non-toxicity. Moreover, the incorporation of certain environmentally safe nanoparticles has led to the formulation of green nanocutting fluids based on vegetable oils, offering both operational efficiency and environmental safety (wang et al. 2025).

Based on their formulation and intended use, cutting fluids are generally categorized into four primary types. Neat oils, often referred to as straight oils, do not mix with water and are known for offering superior lubricating properties while having minimal cooling capabilities. These are typically applied in slower machining processes where moderate heat is produced. In contrast, soluble oils consist of oil blended with water to create a cloudy emulsion, offering both cooling and lubrication, making them appropriate for a variety of machining operations. Semi-synthetic fluids combine small amounts of oil with synthetic components and water, providing better stability, cleanliness, and cooling than soluble oils. Synthetic fluids, which contain no oil, are water-based solutions with chemical additives, offering superior cooling and are ideal for high-speed machining where heat dissipation is critical.

2.4.1 Study on conventional cutting fluids

Metalworking fluids (MWFs) have significantly contributed to improving machining operations, yet they come with certain limitations. One of the primary concerns is the high expenditure linked to their handling, which includes storage, recycling, and disposal. In the European Union, over 66% of cutting fluids are discarded without being reused (Sarıkaya et al., 2022). The expenses related to storing cutting fluids can be twice as much as the actual machining expenses. Approximately 20% of the production cost of automotive components can be attributed to cutting fluid usage (Jamwal et al., 2022). When considering the total cost of manufacturing, nearly one-fifth is consumed by the usage, handling, and disposal of cutting fluids, whereas the tool cost makes up just 4% (Sen et al., 2021; Souflas et al., 2022). For machining hardened materials, the expense related to MWFs can rise to nearly four times in comparison to machining of conventional materials (Vukelic et al., 2020). Additionally, cutting fluid expenditures contribute significantly to the product's indirect costs (Singh et al., 2020).

A considerable amount of lubricant is typically introduced into the cutting zone during material removal processes. However, as the tool or workpiece rotates, part of the fluid is dispersed away from the cutting area a phenomenon referred to as drag-out. This fluid loss is heavily influenced by the geometry and design of the workpiece. Additionally, metal chips generated during cutting can carry some fluid away, further contributing to material loss. These losses increase the indirect manufacturing expenses (Madanchi et al., 2015; Mulyadi et al., 2015). Reusing degraded cutting fluids introduces further complications, including rusting and oxidation of tools and components. Aged fluids can also lead to increased tool wear, unpleasant smells, and microbial contamination. Recycling used cutting fluids demands substantial water resources for processing. The procedure typically includes separating oil from water and subsequently treating both phases, which requires energy and materials (Sarikaya et al., 2021; Khanna et al., 2022). Water-based cutting fluids are especially prone to microbial activity and thus need biocides to inhibit bacterial and fungal growth. Frequently used antimicrobial agents consist of triazine oxazolone, di-cyclohexylamine, and phenoxyethanol. (Trafny et al., 2015; Özcan et al., 2019).

Operators working on metalworking machines often come into direct contact with cutting fluids during operation. Several studies have reported that prolonged and repeated exposure to these fluids and their decomposition products can result in health issues such as skin disorders, respiratory illnesses, and, in severe cases, cancer (Hannu et al., 2013; Do et al., 2018; Thorneus et al., 2021). When the skin is broken or injured, contact with certain cutting fluid additives like sulfites and chlorine compound can cause irritation to the affected area. To reduce health risks, it is essential for operators to wear appropriate personal protective equipment, including gloves, safety boots, and face masks, while handling or working near these fluids. While traditional metalworking fluids effectively lower cutting temperatures, concerns related to environmental impact, cost inefficiencies, and occupational health risks have driven the industry toward adopting greener and more sustainable machining practices. Numerous studies have indicated that relying heavily on cutting fluids during machining is not a viable approach for sustainable manufacturing. Hence, investigating alternative methods for cooling and lubricating in machining processes is crucial. (Chetan et al., 2015a; Khan et al., 2020; Sankaranarayanan et al., 2021). In the past twenty years, several eco-friendly machining approaches have been explored and analysed by researchers, including dry cutting, solid lubricants, MQL, cryogenic techniques, and heat pipe-assisted machining.

2.4.2 Study on Flood cooling

In conventional machining, cutting fluids are extensively utilized in what is commonly referred to as flood coolant machining. These fluids serve the dual purpose of dissipating heat generated from the cutting zone and providing lubrication at the tool-workpiece interface. This, in turn, enhances tool longevity and improves the surface finish of the machined component. The exceptional surface quality that flood coolant machining is primarily due to its ability to minimize adhesion wear, thereby reducing built-up edge formation (Thakur and Gangopadhyay, 2016; Zhou et al., 2012). Additionally, cutting fluids assist in chip evacuation from the cutting area and offer corrosion protection for both the tool and the workpiece. The commonly used cutting fluids in machining include water-soluble oils, mineral oils, and vegetable-based oils.

The researchers employed a water-based emulsion coolant with a flow rate of 40 liters per minute at a pressure of 5 bar. Their findings highlighted that flood cooling significantly contributed to improving surface finish and extending tool life. Cantero (2013) examined the wear study of tools during finish face turning of Inconel 718, utilizing various insert types under dry and wet cutting environments. The findings revealed that the use of flood coolant significantly enhanced tool longevity. The primary reason for increased flank wear in dry machining was attributed to excessive cutting temperatures, which intensified abrasive wear. Despite its benefits, flood coolant machining presents several sustainability challenges. The high consumption of cutting fluids leads to elevated operational and disposal costs, along with potential health hazards. Furthermore, most cutting fluids contain toxic and non-biodegradable components, raising environmental concerns. Another critical issue is the excessive use of water, which is required not only for preparing cutting fluids but also for compensating for evaporation losses and cleaning machined components after processing (Salman et al., 2018).

2.4.3 MQL technique in machining

Numerous studies have focused on creating eco-conscious cooling and lubrication techniques to replace traditional flood-based machining methods. In a detailed review of various cutting fluid application techniques, Chetan et al. (2015) analyzed the environmental and safety concerns associated with flood coolant usage and assessed multiple sustainable alternatives. These alternatives include MQL, cryogenic cooling, air/gas cooling, and biodegradable cutting fluids. The major findings from their study are as follows:

1) Cryogenic cooling enhances machining efficiency at higher speeds and feed rates; however, it is a costly approach.

2) MQL provides comparable or superior performance to flood cooling while mitigating health, safety, environmental, and maintenance concerns, though the application of compressed air in MQL can lead to mist formation.

3) Compressed air, while a potential substitute, is not highly effective for machining superalloys.

Sharma et al. (2016) reviewed the use of MQL technique in machining and found that it delivers comparable, and sometimes superior, results to flood cooling in terms of surface quality, cutting force reduction, and tool wear. Obikawa et al. (2013) examined the effect of air-jet-assisted cooling on Inconel 718 machining and reported a 20–30% increase in tool life over conventional wet cutting. In an earlier investigation, Obikawa et al. (2008) explored micro-lubrication techniques during the turning of Inconel 718, concluding that oil mist lubrication enhances the stability of the MQL system and boosts overall machining performance.

Courbon et al. (2011) reported that high-pressure jet-assisted (HPJA) machining effectively reduced cutting forces during Inconel 718 machining. Similarly, Tazehkandi et al. (2016) explored the use of spray-mode cutting fluid application with compressed air in turning Inconel 783 using PCBN tools and observed improvements in both cutting forces and surface roughness. While MQL is increasingly considered a practical substitute for conventional flood cooling, choosing an appropriate lubricant remains essential and optimization of flow rate are crucial for achieving desirable outcomes. Sarıkaya and Güllü (2014) conducted multi-response optimization of MQL parameters while machining Haynes 25 alloy. They employed aerosol spray lubrication, where cutting fluid was mixed with air at 6 bar pressure. Results indicated that using vegetable oil at a delivery rate of 180 ml/hr, along with a cutting velocity of 30 m/min, yielded optimal outcomes by minimizing tool degradation and improving the machined surface finish.

Chetan et al. (2016) investigated the application of MQL with a water-soluble oil delivered at 3.5 bar pressure and 180 ml/hr during the machining of Nimonic 90. Their study revealed that compared to dry machining, MQL significantly improved the surface finish and reduced tool wear by half. Sharma et al. (2016) reviewed MQL applications across various materials and machining processes. Their study highlighted that MQL consistently outperforms dry machining and emphasized the need for further research comparing MQL with flood coolant machining. Gupta et al. (2016) investigated the optimization of machining parameters for titanium alloys using MQL conditions (300 ml/hr flow rate, 4 bar air pressure) and found that higher feed rates led to an increase in tangential force, tool wear, and surface roughness, while cutting speed significantly influenced all three outcomes.

Tebaldo et al. (2017) assessed different cutting fluid application methods for turning Inconel 718, using compressed air at 6 bar pressure in minimum quantity cooling lubrication (MQCL). Their findings indicated that MQCL delivered tool wear and surface finish results similar to those obtained through conventional flood cooling. Musavi et al. (2018) compared MQL and MQCL methods with dry and wet machining while turning A286 alloy. Their study used compressed air at 6 bar pressure in MQL and liquid nitrogen for cooling. The findings showed that both MQL and MQCL approaches offered notable enhancements in surface quality and decreasing tool wear over conventional methods. Magnesium alloy turning was studied by Viswanathana et al. (2018) under dry and MQL conditions, utilizing a 6 bar compressed air system to generate a lubricant mist. Their results indicated that MQL provided superior performance by minimizing tool degradation, lowering surface roughness, reducing cutting forces, and effectively decreasing machining temperature.

2.4.4 Performance of vegetable oils based cutting fluids with MQL technique

Numerous investigations have demonstrated that cutting fluids derived from vegetable sources offer enhanced lubrication and are more readily biodegradable than those based on mineral oils. As noted by Lue et al. (2024), vegetable oils are mainly composed of triglycerides and fatty acids, which possess chemically active polar groups such as esters (COOR) and carboxyls (COOH). Debnath et al. (2014) highlighted that the superior lubricating ability of vegetable oils stems from their polar functional groups, which assist in developing a molecular film over the workpiece surface, thereby reducing both friction and tool deterioration. Belluco and Chiffre (2004) explored how five types of vegetable oils performed during the drilling of AISI 316L stainless steel. They assessed aspects such as chip formation, the longevity of cutting tools, and the forces involved in cutting. Their research revealed that the use of vegetable oils as lubricants could increase tool life by as much as 177%.

Ojolo et al. (2008) carried out experimental investigations to assess the impact of various plant-based oils including coconut, palm kernel, shea butter, groundnut, and palm oils on cutting force behavior during machining of aluminum, copper, and mild steel. Their results showed that vegetable oils positively influenced turning operations, though cutting force varied depending on the material to be cut. When turning AISI 9310 steel with uncoated carbide inserts, Khan et al. (2009) investigated the usage of lubricants based on vegetable oil in a MQLmethod. The investigation focused on tool wear behavior, surface finish, chip characteristics, and thermal conditions. They examined the results of MQL using both traditional flood cooling and dry cutting. Higher cutting speeds and feed rates were shown to raise the temperature, but they also found that temperature decrease had no impact on the type of chip creation. MQL-assisted machining, on the other hand, performed better than flood and dry cooling.

Xavior and Adithan (2009) studied to assess the influence of bio-based lubricants on surface finish and tool degradation during the machining of AISI 304 stainless steel. The tests were performed using carbide cutting inserts, and the lubricants examined included coconut oil, a conventional emulsion, and a non-water-soluble oil. Results indicated that coconut oil notably enhanced machining outcomes, whereas the soluble oil had little effect on both tool degradation and surface roughness. In a separate study, Kuram et al. (2010) investigated the drilling of AISI 304 stainless steel using a high-speed steel (HSS-E) drill to assess the impact of various cutting fluids on thrust force and surface roughness. Their analysis indicated that a spindle speed of 720 rpm produced favorable outcomes with the use of commercial oil led to the lowest machining response metrics.

Mia et al. (2018) explored the effectiveness of using olive oil as an eco-friendly lubricant in MQL turning of hardened steel. Their investigation was centered on evaluating key outcomes such as surface quality and thermal conditions in the machining. When compared to other cooling techniques, their findings showed that MQL with olive oil represented the most environmentally friendly machining solution. A later investigation examined the performance of various plant-based oils as lubricants during the grinding process of GH4169 alloy. Maize oil, rapeseed oil, soybean oil, sunflower oil, peanut oil, palm oil, and castor oil were experimentally ranked in order of increasing efficacy based on their lubricating capabilities. The analysis took into account, specific energy, grinding force friction coefficient, grinding ratio, and the appearance of the machined surface, highlighting how different vegetable oils can impact grinding efficiency in metalworking processes.

2.4.5 Performance of nano fluids with MQL techniques

Numerous studies have shown that employing nanofluids with MQL in operations like turning, drilling, and milling offers notable advantages. The presence of nanoscale additives dispersed within the lubricant base oil enhances thermal conductivity, enabling more efficient removal of heat from the machining area. The improved durability of cutting tools when using MQL nanofluids is credited to These nano-additives are serving as barriers at the tool-workpiece interface, thereby reducing the friction coefficient. As a result, this approach helps to decrease cutting forces, tool wear, and the roughness of the machined surface. MQL achieves optimal performance when combined with nanofluids due to several key lubrication mechanisms. Nano-sized spherical particles facilitate a rolling motion between the contact surfaces, efficiently combining rolling and sliding friction to create sliding friction. A protective surface film forms as the nanoparticles interact with the friction pairs, reducing wear and improving surface integrity. Over time, tribo-films develop through the deposition of nanoparticles on the contact surfaces, creating a self-repairing effect that compensates for material loss. Furthermore, the nanoparticles help distribute compressive forces more evenly across the interface, thereby mitigating stress concentrations caused by high contact pressure.

Hegab et al. (2018) explored the working mechanism of MQL-nanofluids. Their findings indicate that these nanofluids are atomized using an MQL device in conjunction with compressed air, forming a fine mist encapsulated in an oil layer. This mist effectively penetrates the tool-workpiece interface, generating a tribo-film that plays a crucial role in reducing cutting heat and minimizing friction. These enhancements significantly improve cooling and lubrication performance while preserving tool hardness over extended periods. As a result, MQL-nanofluids outperform conventional MQL systems that lack nano-additives in terms of tool wear behavior. Moreover, researchers have examined the abrasive effects associated with MQL-nanofluids. When nano-additive concentrations are too high, excessive particles collide with surface asperities, increasing cutting forces and accelerating wear. Consequently, excessive nano-additives can lead to higher flank wear, negatively impacting surface quality. However, increasing nano-additive concentration also enhances lubrication by introducing more particles into the workpiece interface, reducing friction as the mist acts as a spacer in the cutting zone. Therefore, it is essential to carefully optimize nano-additive concentration to achieve an optimal balance between lubrication, wear reduction, and machining efficiency.

Mandal et al. (2012) examined how different concentrations of nanoparticles influence MQL-assisted grinding performance. They utilized AISI 52100 as the substrate and formulated nanofluids with Al_2O_3 particles measuring 40 nm and 80 nm in diameter, incorporating them at volume ratios of 1%, 3%, and 5%. The findings demonstrated that increasing the nanoparticle content resulted in less favorable machining results relative to using lower concentrations. In a related investigation, Setti et al. (2012) explored how incorporating aluminum oxide (Al_2O_3) and copper oxide (CuO) nanoparticles influenced the grinding behavior of Ti-6Al-4V alloy. The results show that using Al_2O_3 nanoparticles led to the most significant reduction in friction levels compared to machining under dry conditions, conventional wet methods, and nanofluids containing CuO particles. In related work, Mao et al. (2013) and Mao et al. (2014) studied the application of nanofluids containing Al_2O_3 particles for grinding AISI 52100. They discovered that the addition of Al_2O_3 nanoparticles to deionized water effectively reduced grinding resistance and minimized friction levels. Nevertheless, the use of larger particulate sizes resulted in poorer surface quality. In another study, Sharma et al. (2016) developed nanofluids by incorporating 1% volume fraction of Al_2O_3 nanoparticles into both vegetable oil and a water-based emulsion, using these mixtures as lubricants under MQL conditions while turning AISI 1040 steel. In comparison to traditional flood cooling, their experiments showed improved machining performance, with a 25.5% reduction in surface roughness, a 5.27% decline in tool wear, and a 28% drop in cutting forces.

SiO_2 and graphene nanoparticles are chosen for vegetable oil-based nanofluids due to their ability to significantly improve the thermal conductivity, viscosity, and tribological properties of the base oil. The addition of these nanoparticles leads to superior heat dissipation and lubrication at the tool-workpiece interface, which is critical for efficient machining operations. Graphene nanoparticles significantly enhance the fluid's thermal conductivity and wetting ability, whereas SiO_2 nanoparticles improve the nanofluid's flow characteristics and stability, leading to superior cooling and lubrication efficiency in machining operations. (Baldin et al. 2022).

When used together, SiO_2 and graphene nanoparticles exhibit synergistic effects that further improve the cutting fluid's performance. The hybrid nanofluid forms a composite protective film at the interface, where graphene provides a self-lubricating layer and SiO_2 forms a tribo-thin layer, reducing direct contact and smoothing surface asperities. This synergy leads to reduced friction, lower tool wear, and improved surface finish compared to using either nanoparticle alone (Huang et al. 2018). The ideal blend of SiO_2 and graphene nanoparticles can be customized to enhance particular characteristics like thermal conductivity or wettability, based on the demands of the machining process. Nanofluids derived from vegetable oils containing both SiO_2 and graphene have shown outstanding results in decreasing cutting forces, improving surface finish, and reducing tool wear, as well as lowering the temperature during cutting. These advancements help extend tool longevity and achieve higher quality surfaces on machined components (Baldin et al. 2022, Pal et al. 2020). Additionally, using vegetable oils as the base fluid supports sustainability goals by reducing reliance on synthetic oils and minimizing environmental impact.

Sarhan et al. (2012) studied the application of SiO_2-based nanolubricants in milling processes and found that they notably lowered both cutting forces and specific energy requirements compared to conventional mineral oils. In another investigation, Sayuti et al. (2014) examined the effectiveness of SiO_2-enhanced nanofluids in machining AISI 4140 steel. They reported that optimal surface quality was achieved using a 300° MQL nozzle angle, 0.5% SiO_2 nanoparticle content, and low compressed air pressure. In contrast, the lowest rate of tool wear occurred using a 600° nozzle angle, the same nanoparticle concentration, and pressure of 2 bar. Similarly, Ooi et al. (2015) conducted milling operations on Al-6061-T6 alloy utilizing SiO_2-based nanofluids and implemented a fuzzy logic framework to enhance the accuracy of predicting experimental outcomes, achieving a correlation accuracy ranging between 91% and 98% with real-world data. Meanwhile, You and Gao (2009) studied carbon nanotube (CNT) integration in nano-machining applications, highlighting that grinding wheels embedded with CNTs yielded favorable outcomes and could function effectively as nano-scale cutting abrasives.

Rao et al. (2011) evaluated the effect of carbon nanotube (CNT) additives during turning processes. They varied the concentration of CNTs in the MQL medium and observed that incorporating 2% CNT yielded the best results in terms of minimizing both tool wear and temperature of cutting zone. In another study, Prabhu and Vinayagam (2012) utilized atomic force microscopy (AFM) to examine the grinding behavior of AISI D3 steel. They applied Taguchi's design approach for optimizing machining parameters and used regression techniques to forecast performance metrics. Their findings revealed that CNT-based nanofluids substantially decreased the occurrence of micro-cracks and improved surface finish compared to dry grinding conditions. Graphene nanoplatelets (GnP) have emerged as a promising additive among various nanoparticles, largely due to their excellent heat conductivity and superior lubricating characteristics. Recently, a growing body of research has focused on the application of GnP in MQL-based machining. For example, Huang et al. (2006) performed experiments showing that the addition of GnP to paraffin oil significantly enhanced its resistance to wear. In a comparable investigation, Marcon et al. (2010) explored the application of cutting fluids infused with graphene nanoplatelets (GnP) during micro-milling processes and observed a marked reduction in tool degradation and thermal buildup in the machining area

Alberts et al. (2009) explored the efficiency of GnP in surface grinding processes, concluding that GnP-infused cutting fluids were particularly advantageous for grinding D-2 steel. In another study, Padmini et al. (2015) developed lubricants by incorporating both nano- and microparticles of boric acid and MoS2 into sesame and coconut oils. Their study revealed that nanofluids provided superior performance compared to microfluids in reducing cutting forces, surface finish irregularities, and machining temperatures. In another investigation, Zhang et al. (2015) added MoS_2 nanoparticles to palm, soybean, and rapeseed oils and used these nanofluids during MQL-assisted grinding. Their findings indicated that the combination of MoS_2 with soybean oil offered the most effective lubrication for grinding operations. Expanding on this research, Padmini et al. (2016) evaluated how MoS_2 nanoparticles performed when blended with sesame, coconut, and canola oils for machining AISI 1040 steel. The study indicated that incorporating 0.5% molybdenum disulfide (MoS_2) into coconut oil delivered the best performance, with significant reductions in cutting force (37%), temperature (21%), tool wear (44%), and surface roughness (39%) relative to dry machining conditions. In another investigation, Nam et al. (2011) explored the application of diamond nanoparticles (30 nm size) suspended in vegetable and paraffin oils for lubricating during Al-6061 machining. The tests demonstrated significant torque reduction for both nanofluids, though the paraffin-based variant delivered better lubrication than the vegetable oil formulation. Subsequently, Nam et al. (2013) concluded that larger diamond nanoparticles were not suitable for effective lubrication in drilling operations.

Singh et al. (2021) attempted to inhance the lubrication ability of jatropha oil by dispersing varying concentrations of Al_2O_3 and ZrO nanoparticles. They observed that the most preferred viscosity was obtained by dispersing 1 wt.% of Al_2O_3. However, the best thermal properties were obtained by dispersing 0.5 wt.% of ZrO. Thus, there was no composition that provided the best properties in all aspects. The authors also conducted a life cycle assessment and found that jatropha-based nanofluids have less effect on the environment. Further attempts have been made to improve the lubrication effect of jatropha oil by adding different combinations of nanoparticles as well as ionic liquids. Sirin and Kivak (2021) investigated the effect of adding boron nitride, graphite and MoS_2 nanoparticles in a vegetable oil named Cuttex SYN 5. They highlighted that the size of nanoparticles for preparing the nanofluid should be less than 100 nm.

Koshy et al. (2015) explored the frictional behavior and wear characteristics of coconut oil dispersed with different wt.% molybdenum disulfide (MoS_2) nanoparticles. Experiments conducted across a temperature spectrum of 30 °C to 120 °C. The inclusion of nanoparticles led to noticeable increases in both frictional performance and material wear behavior. In another study, Makhesana et al. (2022) investigated the lubrication ability of canola oil by adding different weight concentrations of MoS_2, viz. 0.5%, 1% and 1.5%. The use of nano-lubricants has been associated with decreased energy consumption during machining operations. In a recent study, Ross et al. (2023) examined the effectiveness of nano-enhanced coolant on the machining of nickel alloys. Their research involved suspending multi-walled carbon nanotubes (MWCNTs) at various weight percentages 0.2, 0.4, 0.6, 0.8, and 1 in soybean oil to evaluate performance. For comparison, experiments in other cooling environments, viz., dry, minimum quantity lubrication and cryogenic carbon dioxide were conducted. It was noted that the wettability, thermal conductivity, and viscosity are improved by the addition of MWCNT particles in soybean oil; the surface finish also improved. Cui et al. (2022) prepared a nanofluid lubricant from waste soybean oil. The nanofluid was utilized for MQL in machining process at a reduced cost.

Ravuri et al. (2016) utilized a grinding wheel embedded with graphene nanoplatelets (GnP) for machining Inconel 718, noting substantial improvements in the grinding wheel's durability and the surface integrity of the workpiece. Likewise, Pavan et al. (2019) performed experimental research in an MQL-assisted grinding environment and found that adding 0.3% GnP to the base lubricant significantly enhanced the surface finish of Inconel 718. Additionally, Kwon and Drazel (2015) secured a patent for an MQL system based on exfoliated graphene nanoplatelets (XGnP), showcasing the advantages of GnP in reducing tool chipping and increasing tool life. Wang et al. (2016) evaluated the lubricating efficiency of various nanoparticles, discovering that Al_2O_3 nanoparticles provided superior lubrication compared to others. Their tests on GH4169 alloy using palm oil as the MQL base fluid established a ranking of nanoparticle performance as follows: $ZrO_2 < CNT < ND < MoS_2 <$

$SiO_2 < Al2O_3$. Of the various compounds, nanoparticles such as Al_2O_3, MoS_2, and SiO_2 are considered more suitable as additives in nanofluids.

2.5 Influence of Parameters on Machining Performance

Cutting parameters including spindle speed, material feed rate, and depth of cut significantly influence machining performance in determining overall machining performance. These parameters directly influence surface finish, MRR, tool wear, cutting temperature, and dimensional accuracy. Proper selection and optimization of machining parameters enhance productivity, extend tool life, and ensure superior quality of the machined components.

2.5.1 Effect of machining parameters on surface roughness

Dimensional accuracy and surface texture play a crucial role in defining the quality of a machined part. Obtaining the required surface finish depends on the proper choice of lubrication conditions, cutting fluids, machining settings, and tooling. Surface finish is influenced by many variables, including the tool's travel speed, MRR, amount of material cut in one pass, the kind of lubricant used, the curvature of the tool tip, the workpiece's hardness, and the presence of irregular material buildup on the tool. When no lubricant is used during dry machining, increased contact between the tool and workpiece generates more heat due to higher friction in the cutting area. This elevated temperature accelerates abrasion, diffusion, and oxidation, ultimately degrading the surface finish. Additionally, significant heat generation within the workpiece induces plastic deformation, leading to metallurgical damage on its outer layer. Cetin et al. (2019) examined how various cutting parameters and cutting fluids influence surface roughness during the machining of AISI 304L stainless steel. Their analysis showed that feed rate had the largest effect, accounting for 97.40% of the variation in surface roughness, while spindle speed, cutting depth, and cutting fluid contributed 0.23%, 0.84%, and 0.69%, respectively.

Machining Titanium Grade-2 is difficult due to its strong toughness and poor ability to conduct heat. At cutting speeds between 200 and 250 m/min, the alloy generates substantial heat, which is mostly absorbed by the workpiece due to its limited ability to dissipate heat. This increase in temperature leads to material softening, which in turn promotes better material flow and elastic deformation. Consequently, the inherent motion within the surface layers helps to lower resistance at the contact zone between the tool and the workpiece, enhancing the quality of the machined surface and reducing tool degradation.

However, when the cutting speed is elevated from 250 to 300 m/min, the heat generated becomes excessive and exceeds what the cutting fluid can dissipate. The heat makes the material softer, which causes more bending and faster tool damage, leading to a rougher surface. Kumar et al. (2021) studie the turning of Inconel 718 using dry cutting, flood cooling, and MQL techniques. They found that as cutting speed increased to 70 m/min, surface finish improved. The material became softer as the temperature rose at faster rates, making it easier to cut and resulting in a finer surface texture. However, the cutting zone's temperature rises over tolerable levels, when the cutting speed surpasses 75 m/min across all lubrication conditions due to poor heat dissipation of Inconel 718.

Based on metal cutting kinematics, raising the feed rate results in increased surface roughness. A higher feed enlarges the cut's cross-sectional area, which increases the volume of material deformation and adversely affects surface finish. Observations reveal that when the feed rate is between 0.05 and 0.1 mm/rev, surface roughness rises gradually with feed increments. However, once the feed exceeds 0.15 mm/rev, there is a rapid increase in surface roughness. The application of nanofluid in an MQL environment enhances machining efficiency. A thin lubricating layer forms between the tool and workpiece, promoting faster heat dissipation due to the nanofluid's superior thermal conductivity. Furthermore, nanoparticles suspended in the fluid act like tiny rollers, lowering interfacial friction between tool and material. As a result, employing nanofluids in a MQL system enhances surface quality, reduces tool degradation, and lowers cutting forces. Gupta et al. (2021) examined the effect of high-speed turning of Inconel 800 superalloy under MQL using sunflower oil-based nanofluids containing Al_2O_3, MoS_2, and graphite nanoparticles. The incorporation of nanofluids led to improved surface quality with reduced surface roughness during high-speed machining. While all three nanofluids exhibited comparable performance, MoS_2 and graphite demonstrated superior results due to their higher thermal conductivity compared to the Al_2O_3 based nanofluid.

An increase in cutting speed typically leads to higher surface roughness. This is primarily due to intensified machine tool vibrations that occur at elevated speeds, particularly in the range of 200 to 300 m/min, during the machining of superalloy. Consequently, at greater feed rates, surface finish deteriorates due to the combined effects of elevated friction and tool damage. Velmurugan et al. (2019) performed a comparative analysis of surface roughness while turning Inconel X-750 and Waspaloy alloys. A total of nine turning experiments were carried out on both materials. It was observed that while machining Inconel X-750, excessive frictional rubbing occurred, causing intermittent machining. Consequently, groove formation and built-up edges were noticed, leading to a decline in surface quality. For both materials, higher feed rates resulted in greater contact of tool and workpiece interface, which produced more pronounced surface irregularities and diminished surface finish quality. This demonstrates that feed rate is a key factor influencing surface roughness. Conversely, the surface finish increased with increasing cutting speeds due to thermal softening effects. For both materials, surface roughness increased with higher feed rates and decreased as cutting speed rose. Optimal surface finish was obtained at elevated cutting speeds combined with reduced feed rates. The measured surface roughness for Inconel X-750 varied from 0.76 to 1.56 μm, while for Waspaloy, it ranged between 0.56 and 1.01 μm.

Tazehkandi et al. (2016) explored the machining performance of Inconel X-750 during turning operations, employing MQL with eco-friendly vegetable-based oil. Due to the alloy's poor machinability and low thermal conductivity, heat tends to accumulate during machining. This heat buildup softens the material and causes elastic deformation, which decreases rubbing action between the tool and the surface, leading to lower cutting forces and improved surface quality. However, further increases in cutting speed generated excessive heat beyond the fluid's cooling capability, resulting in plastic deformation of the workpiece. This caused an increase in cutting forces and a decline in surface quality. Additionally, feed rate played a key role in surface roughness, as higher feed increased friction at the tool-workpiece interface, raising temperatures. This promoted the formation of BUE that scratched the surface, creating grooves and further reducing surface quality while increasing cutting forces. Moreover, a larger cutting depth (d_c) improved the material removal rate; although it increased friction and cutting forces, its influence on surface roughness was relatively minor.

Alaattin et al. (2021) examined how well carbide tools coated with PVD TiAlN/TiN performed when Inconel X-750 was dry turned. Cutting speeds ranging from 40 to 100 m/min and feed rates between 0.1 and 0.2 mm/rev were used in the testing. The results indicated that moderate cutting speeds (40–60 m/min) paired with lower feed rates consistently produced smoother surfaces. The optimal surface finish (R_a = 0.87 μm) was recorded at a speed of 60 m/min and a feed rate of 0.1 mm/rev. Conversely, when speeds surpassed 60 m/min and the feed rate reached 0.2 mm/rev, surface finish deteriorated. The increase in temperature at elevated speeds contributed to BUE formation, which modified the tool edge and negatively affected surface integrity. The findings show that feed rate mostly determines surface roughness, with cutting speed and depth of cut having relatively less of an impact. Superior surface finish is achieved by combining a reduced feed rate with a greater cutting speed within a specific range. However, the depth of cut has very little effect on surface finish. Moreover, applying cutting fluids during machining is essential, as it helps lower friction and temperature at the contact point between the tool and the workpiece, ultimately improving surface finish and extending the lifespan of the cutting tool.

2.5.2 Effect of machining parameters on tool wear

Superalloys based on nickel have a high chemical affinity for different types of cutting tools, which causes an adhesive coating to form and speeds up diffusion wear. Furthermore, the adhesive wear mechanism is exacerbated by their low thermal conductivity, which raises cutting zone temperatures (Hao et al. 2019). Using ceramic cutting inserts and tungsten carbide tools covered with TiN + AlCrN, Charnnarong et al. (2018) assessed the turning performance of AISI 4140 steel at cutting speeds between 150 and 220 m/min. Their findings indicated that cutting speed, insert type, and their interaction significantly influenced flank wear width at a significance level of 0.05. Electron microscopy analysis revealed that Al_2O_3 + TiC inserts experienced mild abrasion through microploughing, microcutting, and microchipping, whereas TiN + AlCrN-coated WC inserts exhibited severe abrasive wear due to direct cutting and abrasion.

During the machining of Inconel X-750 with carbide inserts coated with PVD TiAlN-TiN, material adhesion and the formation of BUE were observed under almost all cutting conditions. Tool wear analysis showed that factors including material removal depth, spindle speed, and feed rate had a significant impact (Şirin et al., 2021). Changes in feed rate had a greater impact on the degree of tool degradation than changes in cutting velocity when the cutting depth remained constant. Energy Dispersive X-ray (EDX) analysis verified the presence of alloying elements like nickel and chromium originating from the workpiece. During dry machining of Inconel X-750 at low feed rates, the tool wear progressed through several mechanisms, such as adhesion, formation of built-up edges, chipping, mechanical fracture, and diffusion wear. Additionally, at a cutting velocity of 75 m/min, alongside a higher feed of 0.125 mm/rev and a steady material removal depth of 0.5 mm, a consistent thermal barrier composed of Al_2O_3 was formed during the machining process. This protective oxide layer greatly improved the tool's lifespan under these conditions.

Xavior and Adithan (2009) evaluated how coconut oil-based lubrication, obtained from plant sources, influenced the machining behavior of AISI 304 stainless steel when processed with carbide inserts. Their results indicated that coconut oil outperformed conventional mineral oils in limiting tool wear. Straight cutting oil, soluble oil, and coconut oil were compared under the same conditions (0.5 mm depth of cut, 0.2, 0.25, and 0.28 mm/rev). The results showed that coconut oil significantly reduced tool damage. Kishawy et al. (2019) investigated tool wear behavior while turning Ti-6Al-4V using two unlike cutting fluids under an MQL conditions. When machining with a conventional MQL fluid (without nanoparticles), extreme abrasive and adhesive wear was observed on the tool's flank and crater surfaces due to intensed rubbing against the workpiece and elevated cutting zone temperatures. However, when machining with an MQL based nanofluid, minimal abrasion and adhesion occurred due to the nanofluid's superior lubrication and heat dissipation properties. An increase in nanoparticle concentration facilitates their penetration into the chip-workpiece interface, where they settle into micro-grooves, creating a layer of protection on the exterior. Combining a feed rate of 0.2 mm/rev, a cutting speed of 120 m/min, and a concentration of 4 weight percent nanoparticles produced the least amount of flank wear.

Stephenson et al. (2014) examined the external turning (rough machining) of unaged Inconel X-750 (Combustor Housing) by replacing conventional water-based flood coolant with supercritical CO_2 ($scCO_2$) as an MQL lubricant. The experiments were carried out on a horizontal turret lathe using coated carbide tools, with soybean oil serving as the lubricant in the $scCO_2$ system. The material removal rate (MRR) exhibited an improvement of 25% and 40% when employing $scCO_2$ MQL. Although notch wear which typically restricts tool life under conventional conditions was significantly mitigated with $scCO_2$ MQL, an increased tendency for crater wear and chip hammering was observed.

Hegab et al. (2018) investigated tool wear behavior during turning using three different fluids: Multi-Walled Carbon Nanotubes (MWCNT)-based nanofluid, Al_2O_3-based nanofluid, and a fluid without nanoparticles. Compared to the fluid without nano-additives, both nanofluids demonstrated superior performance. When utilizing a nanofluid with 2% MWCNT at a feed rate of 0.3 mm/rev and a cutting speed of 30 m/min, the least amount of flank wear was seen. Likewise, a constant cutting speed and a feed rate of 0.4 mm/rev with a nanofluid containing 4 weight percent Al_2O_3 reduced tool degradation. Conversely, the highest level of wear was noted in trial 8, performed at 30 m/min with the same feed rate, where the lubricant lacked any nanoparticle additives.

The enhanced wetting qualities, efficient cooling, and superior lubrication offered by nanofluids are responsible for the noticeable reduction in tool wear. The inclusion of nanoparticles improves heat conduction, convection, and fluid wettability, thereby optimizing the cutting environment. A comparative analysis of two nanoparticle types, MWCNTs and Al_2O_3, in base oil revealed that MWCNTs provided superior machining efficiency in terms of tool wear reduction. This was attributed to exceptional cooling and lubrication capabilities of MWCNTs, which improved the bonding at the tool-workpiece interface. MWCNTs possess a remarkably high thermal conductivity of 3000 W/m·K, significantly surpassing the 35 W/m·K conductivity of Al_2O_3 nanoparticles. Additionally, MWCNT-based nanofluid demonstrated higher zeta potential (strong repulsion between particles, preventing them from aggregating), values compared to Al_2O_3-based nanofluid, further enhancing its stability and tribological properties.

2.5.3 Effect of machining parameters on cutting force

Feed rate, spindle speed, and depth of cut are some of the process variables that have a significant impact on cutting force during machining. Because of increased material removal and tool engagement, larger cutting pressures are typically the result of increased feed rate and depth of cut, while higher cutting speeds may reduce cutting force by lowering material strength at elevated temperatures. Gupta et al. (2021) conducted high-speed turning experiments on Titanium (Grade-2) alloy using CBN tool inserts under an MQL environment. The study revealed that at an initial cutting speed range of 200 to 250 m/min, the cutting force decreased. However, as the speed increased from 250 to 300 m/min, a significant rise in cutting force was observed. Titanium alloy is a hard material characterized by high toughness and a low heat transfer coefficient. Within the lower speed range of 200 to 250 m/min, the limited heat dissipation of the workpiece material caused substantial heat accumulation in the cutting zone. The workpiece material was softened by the heat produced, which improved its flow properties and made it simpler to deform. This reduced friction at the cutting zone and greatly reduced the cutting force. However, the quantity of heat generated was greater than the cutting fluid's capacity to cool as the cutting speed increased further, to between 250 and 300 m/min. This resulted in increased tool wear and plastic deformation of the workpiece.

Wang et al. (2015) found that using Accu-Lube under MQL reduced tool wear and improved surface finish in up milling but was less effective in down milling due to vibrations. Optimal conditions were 160 mm/min speed, 0.2 mm/tooth feed, and 1 mm depth. Liao et al. (2017) reported that higher oil-water ratios and flow rates of BESAL 25 in MQL lowered tool wear and cutting forces, with 60:40 and 60 ml/hr being most effective. Nanoparticle-enhanced fluids have also shown potential for further improvement. Using a SiAlON ceramic tool, Hao et al. (2017) investigated Inconel 718's dry milling performance. The study varied the cutting speed from 50 to 500 m/min while maintaining a fixed feed rate of 0.09 mm/tooth and a constant cutting depth of 1 mm. The findings indicated that cutting forces decreased with increasing cutting speeds. In terms of tool degradation, abrasive wear and layer peeling were the primary wear mechanisms observed at lower speeds between 50 and 200 m/min.

Using TiAlN-coated tools, Kamdani et al. (2019) examined how the radial depth of cut affected the variation in cutting force when dry milling Inconel 718. According to their findings, the cutting force increased as the radial depth increased. Curiously, the effect of cutting speed varied; at a radial depth of 10 mm, increasing the speed from 100 m/min to 200 m/min decreased the cutting force; but, at a depth of 20 mm, higher speeds led to greater forces. Li and Wang (2013) used carbide-inserted milling cutters to test cutting forces during high-speed dry milling of Inconel 718, employing statistical methods to assess force behavior. These findings collectively underline the importance of selecting appropriate machining parameters in dry milling operations. Carou et al. (2020) studied the turning operations for the maintenance and repair of Ti-6Al-4V bars in the aeronautical industry. The primary objective was to restore the components to their original dimensions and functionality while minimizing cutting forces, as they contribute to increased energy consumption and accelerated tool wear. An experimental setup using complete factorial combinations was utilized to analyzing the effects of four key factors: cooling mode, feed rate, spindle speed, and tool type. The results indicated that lower feed rates contributed to reduced cutting forces, while spindle speed had an insignificant impact. The mode of cooling (dry, MQL, and cryogenic) and tool type were identified as critical factors affecting cutting forces. The study concluded that at lower feed rates, productivity could be enhanced by increasing spindle speed, reducing machining time while maintaining minimal cutting forces. Additionally, lower feed rates helped in reducing energy consumption.

The application of a hybrid nanofluid made by dispersing zinc oxide (ZnO) nanoparticles coated in silver into ethylene glycol was investigated by Barewar et al. (2021). During the milling of Inconel 718, this nanofluid was utilized in MQL mode. Different nanoparticle concentrations were tested, and their thermal and physical behaviors were assessed to determine the most efficient formulation. The selected nanofluid, when applied in nanofluid-based MQL (NFMQL), led to a 23.5% and 13.07% improvement in surface finish, and a 15.38% and 8.56% reduction in temperature compared to dry cutting and regular MQL, respectively.

Haq et al. (2021) conducted face milling experiments on Inconel 718 using a nanofluid composed of 0.2% Cu nanoparticles dispersed in water, applied under MQL at different flow rates of 400, 600, and 800 ml/hr. Among the tested conditions, higher flow rates exhibited better lubrication by effectively minimizing friction. The analysis identified depth of cut as the most critical factor influencing machining performance. In a separate study, Hegab et al. (2018) explored the use of rapeseed oil-based nanofluids infused with either multi-walled carbon nanotubes (MWCNTs) or γ-Al2O3 during MQL-assisted turning of Inconel 718, concluding that MWCNTs delivered more favorable outcomes than Al_2O_3 in terms of cutting efficiency. Korkmaz et al. (2021) investigated the machining performance of the Nimonic 80A superalloy in turning operations by analyzing cutting forces through both experimental and finite element simulations. Their findings revealed that cutting and feed forces increased with higher feed rates and depth of cut. However, cutting forces tended to decrease with increasing cutting speed, although ANOVA results suggested that cutting speed had a minimal effect. Additionally, the radial force increased with feed rate, while its variation with depth of cut was relatively minor.

Another study focused on turning Inconel 718 under MQL conditions, where the researchers developed a force model extending Oxley's Predictive Machining Theory (OPMT). This model incorporated Zorev's dual contact zone theory to characterize friction at the chip-tool interface, contrasting with single contact zone theories that assume a uniform coefficient of friction. The dual contact zone theory divides the chip-tool interaction into a sticking region and a sliding region, enabling more accurate friction modeling. A mechanistic approach was formulated to estimate the coefficient of friction within the sliding zone. The cutting force, thrust force, chip thickness, and contact length between the chip and tool were all predicted by this model using an iterative numerical technique that was implemented in MATLAB. Experimental validation showed that the predicted values closely matched the measured results, with average estimation errors of 6.53% for cutting force and 8.3% for thrust force. The study also demonstrated that a significant decrease in the sticking contact length (lp) and the total chip-tool contact length (H) occurred when the flow rate and air pressure were increased. The suggested model optimizes cutting forces and tool performance, laying the groundwork for a broader use of MQL in machining applications.

2.6 Summary of Literature Review

After conducting an extensive review of the literature, several key conclusions have been identified and are summarized as follows:

- Nickel-based superalloys are difficult to machine because of their low thermal conductivity and tendency to harden quickly during cutting. This often leads to cold welding between fragments of the removed material and the cutting tool, resulting in the development of a BUE. Numerous studies have demonstrated that both material adhesion and BUE formation commonly occur under a wide range of cutting conditions when machining Inconel alloys.
- It has been established that machining parameters including cutting speed, feed rate, depth of cut, lubrication method, and tool coating type play a crucial role in influencing machinability characteristics.
- Physical Vapor Deposition (PVD) coatings are widely utilized in finishing operations due to their ability to maintain sharp cutting edges. These coatings are thinner compared to Chemical Vapor Deposition (CVD) coatings, resulting in enhanced sharpness. Additionally, PVD coatings generate compressive stresses, which, in combination with their small layer thickness, strengthen the cutting edge. However, for applications requiring a higher material removal rate, CVD coatings are preferred.
- Studies on different types of single-layer coatings suggest that PVD TiAlN coatings deliver superior performance due to their high hardness and self-lubricating properties. These attributes prevent workpiece material from adhering to the tool surface, ultimately improving surface integrity and extending tool life.
- Multiple studies have shown that employing MQL with vegetable oil-based nanofluids improves machining efficiency over traditional cutting fluids.
- The cutting tool and the workpiece are effectively lubricated by nanofluids. The nanoparticles within the fluid function as rolling elements, transforming sliding friction into rolling friction, thereby reducing contact pressure, friction, and cutting

forces. However, an excessive concentration of nanoparticles can act as abrasive particles, increasing friction, surface roughness, and tool wear, leading to elevated temperatures and higher cutting forces.

2.7 Major Gaps in Literature

A comprehensive review of literature on hard machining, vegetable oil-based lubricants, nano-lubricants, and optimization techniques based on modeling has identified the following research gaps:

- Most existing studies on hard machining have primarily focused on Inconel 718, but there has been limited investigation into its milling with respect to thermal behavior.
- These alloys, despite their industrial significance, remain under-investigated, resulting in a lack of material-specific process optimization. Moreover, although MQL has been extensively studied across various machining processes, its implementation in milling operations particularly for nickel-based superalloys—remains significantly underexplored.
- The bulk of existing research is skewed toward turning and grinding processes, leaving a notable gap in the application of MQL for milling, which involves different chip formation mechanics and thermal dynamics.
- Over time, the focus of cutting fluid research has shifted from merely enhancing machining performance to addressing environmental concerns, such as biodegradability and sustainability. However, a comprehensive comparative study that evaluates the thermal conductivity, rheological behavior, biodegradation rate, and anti-corrosion properties of different lubricant classes—namely mineral oils, nanofluids, and vegetable oil-based green lubricants—is still lacking. Such a study is essential for establishing standardized benchmarks and guiding the development of high-performance eco-friendly lubricants.
- Although nano-lubricants infused with metallic or ceramic nanoparticles have been in use for over a decade, the potential of vegetable oil-based nano-lubricants remains largely untapped. These lubricants possess promising thermophysical and lubricating properties, making them viable alternatives to conventional synthetic fluids. However, the optimization of nanoparticle concentration in these fluids—critical for

balancing cost, dispersion stability, and performance has not been thoroughly investigated, leading to inconsistent results and limited industrial adoption.

- Furthermore, despite the emergence of artificial intelligence (AI) driven modeling and optimization techniques in modern manufacturing, their practical application in metal cutting is still constrained by a high dependency on designer expertise, especially in defining model topology and iteration procedures. There is a pressing need for intelligent, self-learning optimization frameworks that can function independently of manual input, enabling more robust and generalized process control.
- In addition to these gaps, the broader field of sustainable machining also suffers from a lack of integration between hybrid cooling/lubrication techniques and predictive modeling. Hybrid methods such as Cryo-MQL or compressed air with nanofluids remain under-researched, particularly in the context of their combined thermal and tribological effects. Similarly, there is insufficient focus on tool geometry optimization under sustainable machining conditions, which is crucial for reducing cutting forces and improving tool life.
- Finally, while surface roughness is a commonly studied output, long-term functional performance indicators such as fatigue life, residual stress distribution, and corrosion resistance under different lubrication strategies are rarely addressed, leaving a substantial gap in the understanding of real-world component durability.

2.8 Objective of the Thesis

Based on the literature review, the following objectives have been established.

1. Development and characterization of nanoparticles doped bio-cutting fluids.
2. Performance of the milling of Inconel 718 in dry, air cooling, and emulsion environments.
3. Performance of the milling under nano-doped bio-emulsion cutting fluid with MQL Technique.
4. Optimization of process parameters in Dry, Air, and Nano-doped bio cutting fluid.

References

Akhtar, W., Sun, J. and Chen, W., 2016. Effect of machining parameters on surface integrity in high speed milling of super alloy GH4169/Inconel 718. *Materials and Manufacturing Processes*, *31*(5), pp.620-627.

Antonialli, A.Í.S., Carvalho, M.R.D.D. and Diniz, A.E., 2020. On the machinability of the Ni-30 high-temperature iron-based superalloy. *Journal of the Brazilian Society of Mechanical Sciences and Engineering*, *42*, pp.1-11.

Arısoy, Y.M., Guo, C., Kaftanoğlu, B. and Özel, T., 2016. Investigations on microstructural changes in machining of Inconel 100 alloy using face turning experiments and 3D finite element simulations. *International Journal of Mechanical Sciences*, *107*, pp.80-92.

Arunachalam, R.M., Mannan, M.A. and Spowage, A.C., 2004. Surface integrity when machining age hardened Inconel 718 with coated carbide cutting tools. *International Journal of Machine Tools and Manufacture*, *44*(14), pp.1481-1491.

Bayat, M. and Amini, S., 2025. Experimental study on ultrasonic assisted milling effect on blade distortion. *Proceedings of the Institution of Mechanical Engineers, Part B: Journal of Engineering Manufacture*, *239*(4), pp.558-567.

Baldin, V., da Silva, L.R.R., Gelamo, R.V., Iglesias, A.B., da Silva, R.B., Khanna, N. and Rocha Machado, A., 2022. Influence of graphene nanosheets on thermo-physical and tribological properties of sustainable cutting fluids for MQL application in machining processes. *Lubricants*, *10*(8), p.193.

Behera, B.C., Ghosh, S. and Rao, P.V., 2016. Application of nanofluids during minimum quantity lubrication: a case study in turning process. *Tribology International*, *101*, pp.234-246.

Belluco, W. and De Chiffre, L., 2004. Performance evaluation of vegetable-based oils in drilling austenitic stainless steel. *Journal of materials processing technology*, *148*(2), pp.171-176.

Bhise, V.Y. and Jogi, B.F., 2022. Effect of cutting speed and feed on surface roughness in dry turning of Inconel X-750. *Materials Today: Proceedings*, *61*, pp.587-592.

Boxall, P., Huo, M.L., Macky, K. and Winterton, J., 2019. High-involvement work processes and systems: a review of theory, distribution, outcomes, and tensions. *Research in personnel and human resources management*, pp.1-52.

Cantero, J.L., Díaz-Álvarez, J., Miguélez, M.H. and Marín, N.C., 2013. Analysis of tool wear patterns in finishing turning of Inconel 718. *Wear*, *297*(1-2), pp.885-894.

Carou, D., Rubio, E.M., Agustina, B. and Marín, M.M., 2017. Experimental study for the effective and sustainable repair and maintenance of bars made of Ti-6Al-4V alloy. Application to the aeronautic industry. *Journal of Cleaner Production*, *164*, pp.465-475.

Chetan, Ghosh, S. and Rao, P.V., 2016. Environment friendly machining of Ni–Cr–Co based super alloy using different sustainable techniques. *Materials and Manufacturing Processes*, *31*(7), pp.852-859.

Chetan, Narasimhulu, A., Ghosh, S. and Rao, P.V., 2015. Study of tool wear mechanisms and mathematical modeling of flank wear during machining of Ti alloy (Ti6Al4V). *Journal of The Institution of Engineers (India): Series C*, *96*, pp.279-285.

Courbon, C., Sajn, V., Kramar, D., Rech, J., Kosel, F. and Kopac, J., 2011. Investigation of machining performance in high pressure jet assisted turning of Inconel 718: A numerical model. *Journal of Materials Processing Technology*, *211*(11), pp.1834-1851.

Cui, X., Sun, N., Cao, P., Guo, J. and Ming, P., 2022. Preparation and application of sustainable nanofluid lubricant from waste soybean oil and waste serpentine for green intermittent machining process. Journal of Manufacturing Processes, 77, pp.508–524.

Danish, M., Gupta, M.K., Rubaiee, S., Ahmed, A., Sarıkaya, M. and Krolczyk, G.M., 2022. Environmental, technological and economical aspects of cryogenic assisted hard machining operation of inconel 718: A step towards green manufacturing. *Journal of Cleaner Production*, *337*, p.130483.

Davoodi, B. and Eskandari, B., 2015. Tool wear mechanisms and multi-response optimization of tool life and volume of material removed in turning of N-155 iron–nickel-base superalloy using RSM. *Measurement*, *68*, pp.286-294.

Debnath, S., Reddy, M.M. and Yi, Q.S., 2014. Environmental friendly cutting fluids and cooling techniques in machining: a review. *Journal of cleaner production*, *83*, pp.33-47.

Devillez, A., Le Coz, G., Dominiak, S. and Dudzinski, D., 2011. Dry machining of Inconel 718, workpiece surface integrity. *Journal of Materials Processing Technology*, *211*(10), pp.1590-1598.

Ezilarasan, C. and Velayudham, A., 2013. An experimental analysis and measurement of process performances in machining of nimonic C-263 super alloy. *Measurement*, *46*(1), pp.185-199.

Ezilarasan, C. and Velayudham, A., 2014. Theoretical predictions and experimental validations on machining the Nimonic C-263 super alloy. *Simulation Modelling Practice and Theory*, *40*, pp.192-207.

Ezugwu, E.O. and Bonney, J., 2005. Finish machining of nickel-base Inconel 718 alloy with coated carbide tool under conventional and high-pressure coolant supplies. Tribology transactions, 48(1), pp.76-81.

Ezugwu, E.O., Wang, Z.M. and Machado, A.R., 1999. The machinability of nickel-based alloys: a review. *Journal of Materials Processing Technology*, *86*(1-3), pp.1-16.

Fan Y, Hao Z, Lin J, et al. New observations on tool wear mechanism in machining Inconel 718 under water vapor+ air cooling lubrication cutting conditions. Journal of Cleaner Production 2015; 90: 381–387.

Fan, Y., Hao, Z., Zheng, M., Sun, F. and Yang, S., 2013. Study of surface quality in machining nickel-based alloy Inconel 718. *The International Journal of Advanced Manufacturing Technology*, *69*, pp.2659-2667.

Fernandes, G.H.N., Barbosa, L.M.Q., França, P.H.P., Martins, P.S. and Machado, Á.R., 2023. Towards green machining: wear analysis of a novel eco-friendly cooling strategy for Inconel 718. *The International Journal of Advanced Manufacturing Technology*, *129*(1), pp.107-126.

Fernandes, G.H.N., Bazon, V.T., Barbosa, L.M.Q., França, P.H.P., da Silva, M.B., Uddin, M., Martins, P.S. and Machado, Á.R., 2023. Performance comparison between internally cooled tools and flood cooling during grey cast iron turning. *Journal of Manufacturing Processes*, *85*, pp.817-831.

Gajrani, K.K. and Sankar, M.R., 2020. Role of eco-friendly cutting fluids and cooling techniques in machining. *Materials forming, machining and post processing*, pp.159-181.

Ghadiri Zahrani, E. and Sedghi, A., 2014. Experimental investigation of precision turning of Monel K-500 under dry conditions. *The International Journal of Advanced Manufacturing Technology*, *73*, pp.1265-1272.

Gómez-Parra, A., Puerta, F.J., Rosales, E.I., González-Madrigal, J.M. and Marcos, M., 2013. Study of the influence of cutting parameters on the Ultimate Tensile Strength (UTS) of UNS A92024 alloy dry turned bars. *Procedia Engineering*, *63*, pp.796-803.

Gupta, M.K., Song, Q., Liu, Z., Sarikaya, M., Jamil, M., Mia, M., Singla, A.K., Khan, A.M., Khanna, N. and Pimenov, D.Y., 2021. Environment and economic burden of sustainable cooling/lubrication methods in machining of Inconel-800. *Journal of cleaner production*, *287*, p.125074.

Gupta, M.K., Song, Q., Liu, Z., Sarikaya, M., Jamil, M., Mia, M., Khanna, N. and Krolczyk, G.M., 2021. Experimental characterisation of the performance of hybrid cryo-lubrication assisted turning of Ti–6Al–4V alloy. *Tribology International*, *153*, p.106582.

Gupta, M.K., Sood, P.K. and Sharma, V.S., 2016. Machining parameters optimization of titanium alloy using response surface methodology and particle swarm optimization under minimum-quantity lubrication environment. *Materials and Manufacturing Processes*, *31*(13), pp.1671-1682.

Hannu, T., Suuronen, K., Aalto-Korte, K., Alanko, K., Luukkonen, R., Järvelä, M., Jolanki, R. and Jaakkola, M.S., 2013. Occupational respiratory and skin diseases among Finnish machinists: findings of a large clinical study. *International archives of occupational and environmental health*, *86*, pp.189-197.

Hao, X., Li, Y., Zhao, Z. and Liu, C., 2019. Dynamic machining process planning incorporating in-process workpiece deformation data for large-size aircraft structural parts. *International journal of computer integrated manufacturing*, *32*(2), pp.136-147.

Hao, Z., Cui, R., Fan, Y. and Lin, J., 2019. Diffusion mechanism of tools and simulation in nanoscale cutting the Ni–Fe–Cr series of Nickel-based superalloy. *International Journal of Mechanical Sciences*, *150*, pp.625-636.

Hao, Z., Fan, Y., Lin, J., Ji, F. and Liu, X., 2017. New observations on wear mechanism of self-reinforced SiAlON ceramic tool in milling of Inconel 718. *Archives of Civil and Mechanical Engineering*, *17*(3), pp.467-474.

Haq, M.A.U, Hussain, S., Ali, M.A., Farooq, M.U., Mufti, N.A., Pruncu, C.I. and Wasim, A., 2021. Evaluating the effects of nano-fluids based MQL milling of IN718 associated to sustainable productions. *Journal of Cleaner Production*, *310*, p.127463.

Hegab, H., Darras, B. and Kishawy, H.A., 2018. Sustainability assessment of machining with nano-cutting fluids. *Procedia Manufacturing*, *26*, pp.245-254.

Hegab, H., Salem, A., Rahnamayan, S. and Kishawy, H.A., 2021. Analysis, modeling, and multi-objective optimization of machining Inconel 718 with nano-additives based minimum quantity coolant. *Applied Soft Computing*, *108*, p.107416.

Hegab, H., Umer, U., Deiab, I. and Kishawy, H., 2018. Performance evaluation of Ti–6Al–4V machining using nano-cutting fluids under minimum quantity lubrication. *The International Journal of Advanced Manufacturing Technology*, *95*, pp.4229-4241.

Jadam, T., Datta, S. and Masanta, M., 2021. Influence of cutting tool material on machinability of Inconel 718 superalloy. *Machining Science and Technology*, *25*(3), pp.349-397.

Jafarian, F., Amirabadi, H., Sadri, J. and Banooie, H.R., 2014. Simultaneous optimizing residual stress and surface roughness in turning of Inconel718 superalloy. *Materials and Manufacturing Processes*, *29*(3), pp.337-343.

Jamwal, A., Agrawal, R. and Sharma, M., 2022. Deep learning for manufacturing sustainability: Models, applications in Industry 4.0 and implications. *International Journal of Information Management Data Insights*, *2*(2), p.100107.

Kamdani, K., Ashaary, I., Hassan, S. and Lajis, M.A., 2019. The effect of cutting force and tool wear in milling INCONEL 718. In *Journal of Physics: Conference Series* (Vol. 1150, No. 1, p. 012046). IOP Publishing.

Kantharaj, I., Vijay, S.J., Vasanth, X.A., Mohanasundaram, S. and Rai, R.S., 2021. Heat pipe-embedded tooling for sustainable manufacturing. In *Sustainable Manufacturing and Design* (pp. 63-81). Woodhead Publishing.

Kaya, E. and Akyüz, B., 2017. Effects of cutting parameters on machinability characteristics of Ni-based superalloys: a review. *Open Engineering*, *7*(1), pp.330-342.

Khajehzadeh, M. and Razfar, M.R., 2020. Process parameters influence on laser-assisted machining-induced residual stresses. *Materials and Manufacturing Processes*, *35*(15), pp.1680-1689.

Khan, A.M., He, N., Li, L., Zhao, W. and Jamil, M., 2020. Analysis of productivity and machining efficiency in sustainable machining of titanium alloy. *Procedia Manufacturing*, *43*, pp.111-118.

Khan, M.M.A., Mithu, M.A.H. and Dhar, N.R., 2009. Effects of minimum quantity lubrication on turning AISI 9310 alloy steel using vegetable oil-based cutting fluid. *Journal of materials processing Technology*, *209*(15-16), pp.5573-5583.

Khanna, N., Wadhwa, J., Pitroda, A., Shah, P., Schoop, J. and Sarıkaya, M., 2022. Life cycle assessment of environmentally friendly initiatives for sustainable machining: A short review of current knowledge and a case study. *Sustainable Materials and Technologies*, *32*, p.e00413.

Kilincarslan, S.K. and Cetin, M.H., 2020. Improvement of the milling process performance by using cutting fluids prepared with nano-silver and boric acid. *Journal of Manufacturing Processes*, *56*, pp.707-717.

Kishawy, H.A., Hegab, H., Deiab, I. and Eltaggaz, A., 2019. Sustainability assessment during machining Ti-6Al-4V with nano-additives-based minimum quantity lubrication. *Journal of Manufacturing and Materials Processing*, *3*(3), p.61.

Korkmaz, M.E., Gupta, M.K., Boy, M., Yaşar, N., Krolczyk, G.M. and Günay, M., 2021. Influence of duplex jets MQL and nano-MQL cooling system on machining performance of Nimonic 80A. *Journal of Manufacturing Processes*, *69*, pp.112-124.

Koshy, C.P., Rajendrakumar, P.K. and Thottackkad, M.V., 2015. Evaluation of the tribological and thermo-physical properties of coconut oil added with MoS2 nanoparticles at elevated temperatures. Wear, 330, pp.288–308.

Kumar, A. and Kumar, H., 2022. Experimental investigation into internal roundness of Inconel-718 thick cylinder using chemical assisted magnetic abrasive finishing process. *Materials Today: Proceedings*, *48*, pp.1468-1475.

Kuram, E., Ozcelik, B., Demirbas, E. and Sik, E., 2010, July. Effects of the cutting fluid types and cutting parameters on surface roughness and thrust force. In *Proceedings of the world congress on engineering* (Vol. 2, pp. 978-988).

Li, H. and Wang, J., 2013. Assessment of cutting forces in high-speed milling of Inconel 718 considering the dynamic effects. *Proceedings of the Institution of Mechanical Engineers, Part B: Journal of Engineering Manufacture*, *227*(11), pp.1581-1595.

Liao, Y.S., Liao, C.H. and Lin, H.M., 2017. Study of oil-water ratio and flow rate of MQL fluid in high speed milling of Inconel 718. *International Journal of Precision Engineering and Manufacturing*, *18*, pp.257-262.

Liew, P.J., Shaaroni, A., Sidik, N.A.C. and Yan, J., 2017. An overview of current status of cutting fluids and cooling techniques of turning hard steel. *International Journal of Heat and Mass Transfer*, *114*, pp.380-394.

Luo, X., Wu, S., Wang, D., Yun, Y., An, Q. and Li, C., 2024. Sustainable development of cutting fluids: the comprehensive review of vegetable oil. *Journal of Cleaner Production*, *473*, p.143544.

Madanchi, N., Winter, M. and Herrmann, C., 2015. Cutting fluid drag-out and exhaust air in grinding processes: influence on the eco-efficiency. *Procedia CIRP*, *29*, pp.329-334.

Mahesh, K., Philip, J.T., Joshi, S.N. and Kuriachen, B., 2021. Machinability of Inconel 718: A critical review on the impact of cutting temperatures. *Materials and Manufacturing Processes*, *36*(7), pp.753-791.

Makhesana, M.A., Patel, K.M. and Khanna, N., 2022. Analysis of vegetable oil-based nano-lubricant technique for improving machinability of Inconel 690. Journal of Manufacturing Processes, 77, pp.708–721.

Mandal, B., Singh, R., Das, S. and Banerjee, S., 2012. Development of a grinding fluid delivery technique and its performance evaluation. *Materials and Manufacturing Processes*, *27*(4), pp.436-442.

Manimaran G, Anwar S, Rahman, MA, et al. Investigation of surface modification and tool wear on milling Nimonic 80A under hybrid lubrication. Tribology International 2021; 155: 106762.

Mao, C., Huang, Y., Zhou, X., Gan, H., Zhang, J. and Zhou, Z., 2014. The tribological properties of nanofluid used in minimum quantity lubrication grinding. *The International Journal of Advanced Manufacturing Technology*, *71*, pp.1221-1228.

Mao, C., Zhang, J., Huang, Y., Zou, H., Huang, X. and Zhou, Z., 2013. Investigation on the effect of nanofluid parameters on MQL grinding. *Materials and Manufacturing Processes*, *28*(4), pp.436-442.

Memarianpour, M., Niknam, S.A., Turenne, S. and Balazinski, M., 2021. Study of the effects of initial cutting conditions and transition period on ultimate tool life when machining Inconel 718. *Materials*, *14*(3), p.592.

M'Saoubi, R., Larsson, T., Outeiro, J., Guo, Y., Suslov, S., Saldana, C. and Chandrasekar, S., 2012. Surface integrity analysis of machined Inconel 718 over multiple length scales. *CIRP annals*, *61*(1), pp.99-102.

Mia, M., Dey, P. R., Hossain, M. S., Arafat, M. T., Asaduzzaman, M., Shoriat Ullah, M., et al. (2018), "Taguchi S/N based optimization of machining parameters for surface roughness, tool wear and material removal rate in hard turning under MQL cutting condition", *Measurement,* v. *122*, pp. 380-391.

Mia, M., Gupta, M.K., Singh, G., Królczyk, G. and Pimenov, D.Y., 2018. An approach to cleaner production for machining hardened steel using different cooling-lubrication conditions. *Journal of Cleaner Production*, *187*, pp.1069-1081.

Motorcu, A.R., Kuş, A. and Durgun, I., 2014. The evaluation of the effects of control factors on surface roughness in the drilling of Waspaloy superalloy. *Measurement*, *58*, pp.394-408.

Mulyadi, I.H., Balogun, V.A. and Mativenga, P.T., 2015. Environmental performance evaluation of different cutting environments when milling H13 tool steel. *Journal of cleaner production*, *108*, pp.110-120.

Musavi, S.H., Davoodi, B. and Niknam, S.A., 2018. Environmental friendly turning of A286 superalloy. *Journal of Manufacturing Processes*, *32*, pp.734-743.

Nam, J.S., Kim, D.H. and Lee, S.W., 2013, June. A parametric analysis on micro-drilling process with nanofluid minimum quantity lubrication. In *International Manufacturing Science and Engineering Conference* (Vol. 55461, p. V002T04A016). American Society of Mechanical Engineers.

Nam, J.S., Lee, P.H. and Lee, S.W., 2011. Experimental characterization of micro-drilling process using nanofluid minimum quantity lubrication. *International Journal of Machine Tools and Manufacture*, *51*(7-8), pp.649-652.

Nickel, C., 2000. ASM Specialty Handbook□. *ASM International Materials Park, OH*, *22*, pp.44073-0002.

Obikawa, T. and Funai, K., 2013. Air jet assisted machining of Inconel 718 with whisker reinforced ceramic tool. *Key Engineering Materials*, *554*, pp.2079-2084.

Obikawa, T., Kamata, Y., Asano, Y., Nakayama, K. and Otieno, A.W., 2008. Micro-liter lubrication machining of Inconel 718. *International Journal of Machine Tools and Manufacture*, *48*(15), pp.1605-1612.

Ogedengbe, T.S., Okediji, A.P., Yussouf, A.A., Aderoba, O.A., Abiola, O.A., Alabi, I.O. and Alonge, O.I., 2019, December. The effects of heat generation on cutting tool and machined workpiece. In *Journal of Physics: Conference Series* (Vol. 1378, No. 2, p. 022012). IOP Publishing.

Ojolo, S., Amuda, M., Ogunmola, O. and Ononiwu, C., 2008. Experimental determination of the effect of some straight biological oils on cutting force during cylindrical turning. *Matéria (Rio de Janeiro)*, *13*, pp.650-663.

Ooi, M.E., Sayuti, M. and Sarhan, A.A., 2015. Fuzzy logic-based approach to investigate the novel uses of nano suspended lubrication in precise machining of aerospace AL tempered grade 6061. *Journal of Cleaner Production*, *89*, pp.286-295.

Ozcan, B. and Ozturk, I., 2019. Renewable energy consumption-economic growth nexus in emerging countries: A bootstrap panel causality test. *Renewable and Sustainable Energy Reviews*, *104*, pp.30-37.

Padmini, R., Krishna, P.V. and Mohana Rao, G.K., 2015. Performance assessment of micro and nano solid lubricant suspensions in vegetable oils during machining. *Proceedings of the Institution of Mechanical Engineers, Part B: Journal of Engineering Manufacture*, *229*(12), pp.2196-2204.

Padmini, R., Krishna, P.V. and Mohana Rao, G.K., 2016. Experimental evaluation of nano-molybdenum disulphide and nano-boric acid suspensions in vegetable oils as prospective cutting fluids during turning of AISI 1040 steel. *Proceedings of the Institution of Mechanical Engineers, Part J: Journal of Engineering Tribology*, *230*(5), pp.493-505.

Pal, A., Chatha, S.S. and Sidhu, H.S., 2020. Experimental investigation on the performance of MQL drilling of AISI 321 stainless steel using nano-graphene enhanced vegetable-oil-based cutting fluid. *Tribology international*, *151*, p.106508.

Pavan, R.B., Venu Gopal, A., Amrita, M. and Goriparthi, B.K., 2019. Experimental investigation of graphene nanoplatelets–based minimum quantity lubrication in grinding Inconel 718. *Proceedings of the Institution of Mechanical Engineers, Part B: Journal of Engineering Manufacture*, *233*(2), pp.400-410.

Periane Natarajan S, Vaudreuil S, Chibane H, et al. Tool life and surface integrity characteristics in milling of SLM and C&W inconel 718 in dry and MQL condition. The International Journal of Advanced Manufacturing Technology 2022; 121(1): 647–659.

Pervaiz, S., Kannan, S. and Kishawy, H.A., 2018. An extensive review of the water consumption and cutting fluid based sustainability concerns in the metal cutting sector. *Journal of Cleaner Production*, *197*, pp.134-153.

Pimenov, D.Y., da Silva, L.R.R., Machado, A.R., França, P.H.P., Pintaude, G., Unune, D.R., Kuntoğlu, M. and Krolczyk, G.M., 2024. A comprehensive review of machinability of difficult-to-machine alloys with advanced lubricating and cooling techniques. *Tribology International*, p.109677.

Pimenov, D.Y., da Silva, L.R.R., Machado, A.R., França, P.H.P., Pintaude, G., Unune, D.R., Kuntoğlu, M. and Krolczyk, G.M., 2024. A comprehensive review of machinability of difficult-to-machine alloys with advanced lubricating and cooling techniques. Tribology International, p.109677.

Pimenov, D.Y., Gupta, M.K., da Silva, L.R., Kiran, M., Khanna, N. and Krolczyk, G.M., 2022. Application of measurement systems in tool condition monitoring of Milling: A review of measurement science approach. *Measurement*, *199*, p.111503.

Pinheiro, C., Kondo, M.Y., Amaral, S.S., Callisaya, E.S., De Souza, J.V.C., De Sampaio Alves, M.C. and Ribeiro, M.V., 2021. Effect of machining parameters on turning process of Inconel 718. *Materials and Manufacturing Processes*, *36*(12), pp.1421-1437.

Pusavec, F., Deshpande, A., Yang, S., M'Saoubi, R., Kopac, J., Dillon Jr, O.W. and Jawahir, I.S., 2014. Sustainable machining of high temperature Nickel alloy–Inconel 718: part 1–predictive performance models. *Journal of Cleaner Production*, *81*, pp.255-269.

Rakesh, P.R. and Chakradhar, D., 2023. Machining performance comparison of Inconel 625 superalloy under sustainable machining environments. *Journal of Manufacturing Processes*, *85*, pp.742-755.

Ravuri, B.P., Goriparthi, B.K., Revuru, R.S. and Anne, V.G., 2016. Performance evaluation of grinding wheels impregnated with graphene nanoplatelets. *The International Journal of Advanced Manufacturing Technology*, *85*, pp.2235-2245.

Ross NS, Gopinath C, Nagarajan S, et al. Impact of hybrid cooling approach on milling and surface morphological characteristics of Nimonic 80A alloy. Journal of Manufacturing Processes 2022; 73: 428–439.

Ross, N.S., Ganesh, M., Ananth, M.B.J., Kumar, M., Rai, R., Gupta, M.K. and Korkmaz, M.E., 2023. Development and potential use of MWCNT suspended in vegetable oil as a cutting fluid in machining of Monel 400. Journal of Molecular Liquids, 382, p.121853.

Saleem MQ and Mehmood A. Eco-friendly precision turning of superalloy Inconel 718 using MQL based vegetable oils: tool wear and surface integrity evaluation. Journal of Manufacturing Processes 2022; 73: 112–127.

Sankaranarayanan, R. and Krolczyk, G.M., 2021. A comprehensive review on research developments of vegetable-oil based cutting fluids for sustainable machining challenges. *Journal of Manufacturing Processes*, *67*, pp.286-313.

Sarhan, A.A., Sayuti, M. and Hamdi, M., 2012. Reduction of power and lubricant oil consumption in milling process using a new SiO 2 nanolubrication system. *The International Journal of Advanced Manufacturing Technology*, *63*, pp.505-512.

Sarıkaya, M., Gupta, M.K., Tomaz, I., Pimenov, D.Y., Kuntoğlu, M., Khanna, N., Yıldırım, Ç.V. and Krolczyk, G.M., 2021. A state-of-the-art review on tool wear and surface integrity characteristics in machining of superalloys. *CIRP Journal of Manufacturing Science and Technology*, *35*, pp.624-658.

Sarkar S and Datta S. Machining performance of Inconel 718 under dry, MQL, and nanofluid MQL conditions: application of coconut oil (base fluid) and multi-walled carbon nanotubes as additives. Arabian Journal for Science and Engineering 2021; 46: 2371–2395.

Sayuti, M., Sarhan, A.A. and Salem, F., 2014. Novel uses of SiO2 nano-lubrication system in hard turning process of hardened steel AISI4140 for less tool wear, surface roughness and oil consumption. *Journal of Cleaner Production*, *67*, pp.265-276.

Seenath, A.A. and Sarhan, A.A., 2024. A state-of-the-art review on cutting tool materials and coatings in enhancing the tool performance in machining the superior nickel-based superalloys. *Arabian Journal for Science and Engineering*, *49*(8), pp.10203-10236.

Sen, B., Mia, M., Krolczyk, G.M., Mandal, U.K. and Mondal, S.P., 2021. Eco-friendly cutting fluids in minimum quantity lubrication assisted machining: a review on the perception of sustainable manufacturing. *International Journal of Precision Engineering and Manufacturing-Green Technology*, *8*, pp.249-280.

Setti, D., Ghosh, S. and Rao, P.V., 2012. Application of nano cutting fluid under minimum quantity lubrication (MQL) technique to improve grinding of Ti–6Al–4V alloy. In *Proceedings of World Academy of Science, Engineering and Technology* (Vol. 70, pp. 512-516). World Academy of Science, Engineering and Technology.

Sharma, A.K., Tiwari, A.K. and Dixit, A.R., 2016. Characterization of TiO2, Al2O3 and SiO2 nanoparticle based cutting fluids. *Materials Today: Proceedings*, *3*(6), pp.1890-1898.

Sharma, A.K., Tiwari, A.K. and Dixit, A.R., 2016. Effects of Minimum Quantity Lubrication (MQL) in machining processes using conventional and nanofluid based cutting fluids: A comprehensive review. *Journal of cleaner production*, *127*, pp.1-18.

Sharma, A.K., Tiwari, A.K. and Dixit, A.R., 2016. Effects of Minimum Quantity Lubrication (MQL) in machining processes using conventional and nanofluid based cutting fluids: A comprehensive review. *Journal of cleaner production*, *127*, pp.1-18.

Sharman, A.R.C., Hughes, J.I. and Ridgway, K., 2015. The effect of tool nose radius on surface integrity and residual stresses when turning Inconel 718™. *Journal of Materials Processing Technology*, *216*, pp.123-132.

Shokoohi, Y., Paiva, J.M., Fox-Rabinovich, G., Bork, C.A.S. and Veldhuis, S.C., 2019. Evaluation of the superabsorbent coolant as a new approach to semi-dry machining. *The International Journal of Advanced Manufacturing Technology*, *101*, pp.2037-2050.

Shokrani, A., Dhokia, V. and Newman, S.T., 2012. Environmentally conscious machining of difficult-to-machine materials with regard to cutting fluids. *International Journal of machine Tools and manufacture*, *57*, pp.83-101.

Singh, A., Ghosh, S. and Aravindan, S., 2022. State of art for sustainable machining of nickel-based alloys using coated and uncoated tools and machining of high strength materials using surface modified cutting tools. *Tribology International*, *170*, p.107517.

Singh, G., Aggarwal, V. and Singh, S., 2021. Experimental investigations into machining performance of Hastelloy C-276 in different cooling environments. *Materials and Manufacturing Processes*, *36*(15), pp.1789-1799.

Singh, M.P., Chauhan, A.S., Rai, B., Ghoshal, S. and Prinja, S., 2020. Cost of treatment for cervical cancer in India. Asian Pacific journal of cancer prevention: APJCP, 21(9), p.2639.

Singh, R., Sah, N.K. and Sharma, V., 2021. Development and characterization of unitary and hybrid Al_2O_3 and ZrO dispersed Jatropha oil-based nanofluid for cleaner production. Journal of Cleaner Production, 317, p.128365.

Şirin, E. and Şirin, Ş., 2021. Investigation of the performance of ecological cooling/lubrication methods in the milling of AISI 316L stainless steel. *İmalat Teknolojileri ve Uygulamaları*, *2*(1), pp.75-84.

Şirin, Ş. and Kıvak, T., 2021. Effects of hybrid nanofluids on machining performance in MQL-milling of Inconel X-750 superalloy. Journal of Manufacturing Processes, 70, pp.163–176.

Şirin, Ş., Sarıkaya, M., Yıldırım, Ç.V. and Kıvak, T., 2021. Machinability performance of nickel alloy X-750 with SiAlON ceramic cutting tool under dry, MQL and hBN mixed nanofluid-MQL. *Tribology International*, *153*, p.106673.

Şirin, Ş., Yıldırım, Ç.V., Kıvak, T. and Sarıkaya, M., 2021. Performance of cryogenically treated carbide inserts under sustainable cryo-lubrication assisted milling of Inconel X750 alloy. *Sustainable Materials and Technologies*, *29*, p.e00314.

Song, K., Wang, M., Liu, L., Wang, C., Zan, T. and Yang, B., 2020. Intelligent recognition of milling cutter wear state with cutting parameter independence based on deep learning of spindle current clutter signal. *The International Journal of Advanced Manufacturing Technology*, *109*, pp.929-942.

Souflas, T., Bikas, H., Ghassempouri, M., Salmi, A., Atzeni, E., Saboori, A., Brugnetti, I., Valente, A., Mazzucato, F. and Stavropoulos, P., 2022. A comparative study of dry and cryogenic milling for

Directed Energy Deposited IN718 components: Effect on process and part quality. *The International Journal of Advanced Manufacturing Technology*, pp.1-14.

Stephenson, D.A., Skerlos, S.J., King, A.S. and Supekar, S.D., 2014. Rough turning Inconel 750 with supercritical CO2-based minimum quantity lubrication. *Journal of Materials Processing Technology*, *214*(3), pp.673-680.

Tazehkandi, A.H., Shabgard, M., Kiani, G. and Pilehvarian, F., 2016. Investigation of the influences of polycrystalline cubic boron nitride (PCBN) tool on the reduction of cutting fluid consumption and increase of machining parameters range in turning Inconel 783 using spray mode of cutting fluid with compressed air. *Journal of Cleaner Production*, *135*, pp.1637-1649.

Tazehkandi, A.H., Shabgard, M., Pilehvarian, F. and Farshfroush, N., 2017. Experimental investigations of cutting parameters' influence on cutting forces and surface roughness in turning of Inconel alloy X-750 with biodegradable vegetable oil. *Proceedings of the Institution of Mechanical Engineers, Part B: Journal of Engineering Manufacture*, *231*(9), pp.1516-1527.

Tebaldo, V., di Confiengo, G.G. and Faga, M.G., 2017. Sustainability in machining:"Eco-friendly" turning of Inconel 718. Surface characterisation and economic analysis. *Journal of cleaner production*, *140*, pp.1567-1577.

Thakur, A. and Gangopadhyay, S., 2016. Dry machining of nickel-based super alloy as a sustainable alternative using TiN/TiAlN coated tool. *Journal of cleaner production*, *129*, pp.256-268.

Thorneus, E., Graff, P., Bryngelsson, L., Nordenberg, E., Ghafouri, B., Johansson, H. and Fornander, L., 2021. Occupational exposure to metalworking fluid and the effect on health symptoms—An intervention study. *Journal of occupational and environmental medicine*, *63*(10), pp.e667-e672.

Trafny, E.A., Lewandowski, R., Kozłowska, K., Zawistowska-Marciniak, I. and Stępińska, M., 2015. Microbial contamination and biofilms on machines of metal industry using metalworking fluids with or without biocides. *International Biodeterioration & Biodegradation*, *99*, pp.31-38.

Umbrello, D., 2013. The effects of cutting conditions on surface integrity in machining Inconel 718 alloy. *Key Engineering Materials*, *554*, pp.2093-2100.

Vetri Velmurugan, K., Venkatesan, K., Devendiran, S. and Mathew, A.T., 2019. Investigation of parameters for machining a difficult-to-machine superalloy: inconel X-750 and waspaloy. In *Innovative Design, Analysis and Development Practices in Aerospace and Automotive Engineering (I-DAD 2018) Volume 2* (pp. 199-215). Springer Singapore.

Vetri Velmurugan, K., Venkatesan, K., Devendiran, S. and Mathew, A.T., 2019. Investigation of parameters for machining a difficult-to-machine superalloy: inconel X-750 and waspaloy. In *Innovative Design, Analysis and Development Practices in Aerospace and Automotive Engineering (I-DAD 2018) Volume 2* (pp. 199-215). Springer Singapore.

Viswanathan, R., Ramesh, S. and Subburam, V., 2018. Measurement and optimization of performance characteristics in turning of Mg alloy under dry and MQL conditions. Measurement, 120, pp.107-113.

Vukelić, Đ., Šimunović, K., Svalina, I., Tadić, B., Prica, M. and Šimunović, G., 2020. Modelling and prediction of surface roughness in CNC turning process using neural networks. *Tehnički vjesnik*, *27*(6), pp.1923-1930.

Wang, C., Li, K., Chen, M. and Liu, Z., 2015. Evaluation of minimum quantity lubrication effects by cutting force signals in face milling of Inconel 182 overlays. *Journal of Cleaner Production*, *108*, pp.145–157.

Wang, L.J., Wang, Q., Li, J.M. and Chou, K.C., 2016. Dissolution mechanism of Al2O3 in refining slags containing Ce2O3. *Journal of Mining and Metallurgy, Section B: Metallurgy*, *52*(1), pp.35-40.

Wang, W., Lin, H., Zhou, M., Xie, H., Dai, B. and Han, S., 2025. Formulation, characterization, mechanism, and application of vegetable oil-based nano-cutting fluids. *The International Journal of Advanced Manufacturing Technology*, *136*(3), pp.909-923.

Xavior, M.A. and Adithan, M., 2009. Determining the influence of cutting fluids on tool wear and surface roughness during turning of AISI 304 austenitic stainless steel. *Journal of materials processing technology*, *209*(2), pp.900-909.

Xavior, M.A. and Adithan, M., 2009. Determining the influence of cutting fluids on tool wear and surface roughness during turning of AISI 304 austenitic stainless steel. *Journal of materials processing technology*, *209*(2), pp.900-909.

Xavior, M.A. and Adithan, M., 2012. Evaluation of parametric models in predicting the machining performance. *International Journal of Industrial and Systems Engineering*, *11*(4), pp.406-427.

Xavior, M.A. and Adithan, M., 2012. Evaluation of parametric models in predicting the machining performance. *International Journal of Industrial and Systems Engineering*, *11*(4), pp.406-427.

Yang, Y., Guo, S., Si, L., Liu, T., Dai, Y., Yan, C. and Zhang, C., 2021. Investigation of a new water-based cutting fluid for machining of titanium alloys. *Journal of Manufacturing Processes*, *71*, pp.398-406.

You, J. and Gao, Y., 2009. A study of carbon nanotubes as cutting grains for nano machining. *Advanced Materials Research*, *76*, pp.502-507.

Zhang, S.J., To, S., Wang, S.J. and Zhu, Z.W., 2015. A review of surface roughness generation in ultra-precision machining. *International Journal of Machine Tools and Manufacture*, *91*, pp.76-95.

Zhang, W., Zhang, P., Su, Z. and Wei, G., 2015. Synthesis and sensor applications of MoS 2-based nanocomposites. *Nanoscale*, *7*(44), pp.18364-18378.

Zhang, Y. and Xu, X., 2021. Lattice misfit predictions via the Gaussian process regression for Ni-based single crystal superalloys. Metals and Materials International, 27, pp.235-253.

Zheng, Y., Wu, D., Wang, H., Liang, J. and Liu, X., 2023. Machining fixture and deformation control of aero-engine thin-walled casing. *The International Journal of Advanced Manufacturing Technology*, *129*(11), pp.5601-5614.

Zhou, J.M., Bushlya, V. and Stahl, J.E., 2012. An investigation of surface damage in the high speed turning of Inconel 718 with use of whisker reinforced ceramic tools. *Journal of Materials Processing Technology*, *212*(2), pp.372-384.

Zlatanovic, D.L., Eigenbrod, H., Stötzel, M., Bergmann, J.P. and Hildebrand, J., 2023. Effects of cryogenic cooling on machining of acrylonitrile-butadiene rubber. *Journal of Manufacturing Processes*, *90*, pp.429-442.

CHAPTER

3

Materials and experimentation procedure

3.1 Introduction

The present research focuses on the development and application of a vegetable oil-based cutting fluid derived from Putranjiva roxburghii tree. Initially, pure vegetable oil was prepared and subsequently it was enhanced by dispersing two types of nanoparticles viz. silicon dioxide (SiO_2) and graphene using the two-step method to create stable nanofluids. Following the synthesis, a comprehensive characterization of the nanofluids was conducted, evaluating their physical, thermal, optical, rheological, and wettability properties. Additionally, emulsions and nanoemulsions were formulated by blending the oil and water phases and nanoparticles, and their corresponding characteristics were also analyzed. The machinability of Inconel 718 was evaluated using eco-friendly cooling and lubrication methods, comparing dry cutting, air cooling, and bio-based MQL techniques.

The purpose of the study is to evaluate how these cooling techniques affect important cutting performance metrics, including tool wear, surface roughness, cutting temperature, and cutting forces. This chapter is organized into eight sections to systematically present the experimental methodology and related processes. Section 3.2 describes the selection of materials used in the study. Section 3.3 focuses on the development of lubricants formulated for the experiments. The characterization techniques employed for analyzing the lubricants are detailed in Section 3.4. Section 3.5 explains the experimental setup designed to test the performance of the lubricants under various machining conditions. The optimization techniques used to refine the process parameters are discussed in Section 3.6. Section 3.7 presents the materials characterization techniques applied to evaluate the chemical and physical properties of the materials involved. Finally, Section 3.8 provides a summary of the key points covered in this chapter.

3.2 Materials used in the Present Study

The selection of materials in this research was carried out through a systematic approach to ensure compatibility, performance, and sustainability for the milling of Inconel 718. The materials selected include vegetable oil (from Roxburghii seeds), nanoparticles (SiO_2 and graphene), the workpiece material (Inconel 718), and cutting tool materials (TiAlN-coated carbide inserts). Putranjiva roxburghii seeds were selected as the base source for bio-cutting fluid due to their local availability and renewable nature. Seeds were cleaned, dried, and oil was extracted using cold pressing to preserve their natural lubricating properties. The extracted vegetable oil is biodegradable, non-toxic, and suitable for eco-friendly machining applications. Preliminary studies and literature support the oil's good film-forming ability, thermal stability, and oxidative resistance, making it a viable base fluid for Minimum Quantity Lubrication (MQL) techniques.

Silicon dioxide (SiO_2) was chosen for its stable dispersion and ability to improve heat dissipation. It enhances the thermal properties of vegetable oil while maintaining viscosity within MQL compatible limits. Graphene was selected due to its exceptional thermal conductivity, large surface area, and superior lubricity. It improves heat transfer, reduces tool workpiece friction, and enhances overall machining performance. Blending SiO_2 and graphene nanoparticles produces a nanofluid exhibiting enhanced thermal and tribological properties, making it well-suited for environmentally friendly machining applications.

Inconel 718, a nickel-rich superalloy, is commonly employed in the aerospace and power generation industries, and automotive sectors. It is known for low thermal conductivity, high work hardening, and rapid tool wear during machining. Due to its difficult-to-machine nature, Inconel 718 serves as an ideal material for evaluating the performance of bio-based nanofluids under MQL conditions. Its use provides a benchmark for testing advanced cooling and lubrication strategies. The elemental composition of the Inconel 718 workpiece material, as provided by the supplier, is as follows: 52.82% nickel (Ni), 19.00% chromium (Cr), 17.80% iron (Fe), 5.43% niobium (Nb) and tantalum (Ta), 3.04% molybdenum (Mo), 0.98% titanium (Ti), 0.50% aluminium (Al), 0.23% cobalt (Co), 0.08% manganese (Mn), 0.08% silicon (Si), and 0.04% carbon (C).

In this study, Titanium Aluminium Nitride (TiAlN)-coated cemented carbide inserts (Manufacturer: WIDIA, Grade: PA145) were chosen for their superior durability and effectiveness in machining Inconel 718, a challenging-to-machine nickel-based superalloy. The TiAlN coating provides exceptional thermal stability and oxidation resistance, forming a protective layer of aluminium oxide (Al_2O_3) during cutting, which significantly minimizes heat transfer to the cutting edge. This coating layer helps minimize friction and suppress built-up edge formation, while also improving surface quality and prolonging tool lifespan, particularly during dry or minimum quantity lubrication (MQL) machining.

The WIDIA PA145 grade inserts utilize a TiAlN coating deposited via Physical Vapor Deposition (PVD) a technique renowned for producing thin, dense, and strongly adherent coatings, which are particularly beneficial for high-speed and precision machining applications. The typical coating thickness for this grade ranges from 3 to 5 micrometers, offering an optimal balance between mechanical durability and sharp edge retention. In some cases, a very thin adhesion layer (<1 μm), such as titanium or titanium nitride, may be applied beneath the TiAlN layer to enhance coating-to-substrate bonding. This relatively thin PVD coating allows the insert to maintain a sharp cutting edge, a critical feature when performing face milling operations on tough materials like Inconel 718. The coating structure may consist of a monolayer or nano-laminated architecture, with a high aluminium content that further improves cutting performance and tool stability in extreme machining environments.

3.3 Development of Lubricants

The development of vegetable oil-based lubricants has gained significant attention in recent years due to growing concerns over environmental sustainability, occupational safety, and the limitations of conventional petroleum-based fluids. Vegetable oils are considered eco-friendly alternatives because they are biodegradable, renewable, and possess excellent lubricity and high flash points. The process begins with the selection of suitable vegetable oils such as soybean, sunflower, castor, coconut, or non-edible oils like Putranjiva roxburghii. These oils are then refined to improve their thermal stability, oxidation resistance, and flow behaviour. In many cases, nano-additives such as graphene, metal oxides, or bio-based nanoparticles are also incorporated to enhance the thermal conductivity, anti-wear properties, and overall performance of the cutting fluids.

Formulating a stable emulsion is a critical step, which involves mixing the vegetable oil with water and emulsifiers. This mixture is thoroughly homogenized using ultrasonication or high-speed stirring to ensure long-term stability and proper dispersion of additives. Rheological behaviour, viscosity, and droplet size are carefully controlled to meet the requirements of machining applications.

3.3.1 A brief description of Putranjiva roxburghii

Putranjiva roxburghii, commonly known as Putranjiva, is an evergreen tree native to tropical Asia, recognized for its drooping branches, corky bark, and coriaceous leaves. It belongs to the family Putranjivaceae and is native to the Indian subcontinent and Southeast Asia. This medium to large evergreen tree typically grows to a height of 10 to 15 meters and is known for its glossy, green, and lanceolate to ovate leaves arranged alternately on the branches. It has smooth, grayish-brown bark and produces small, yellowish-green flowers that bloom in clusters. The tree bears ellipsoidal drupes, and its seeds are traditionally used in religious prayer beads and Ayurvedic medicine. The tree is valued in traditional medicine, especially in Ayurveda, for its various therapeutic uses. Different parts of the plant including leaves, bark, roots, flowers, fruits, and seeds are used for their medicinal properties, such as treating snake bites, pain, rheumatism, infertility, and skin ailments.

The wood of the Putranjiva tree is dense and hard, suitable for making furniture and tools. Environmentally, it is a beneficial species providing dense shade, helping control urban temperatures, and supporting biodiversity. The leaves and seed oil are particularly prominent in herbal and Unani medicine, with reported antioxidant, antipyretic, and anti-inflammatory activities (Pandey and Fulara 2022; Kumar 2020). It does not need much water, therefore it is good for gardens, fences, and planting beside roads. Notably, the oil extracted from its seeds has shown anti-inflammatory and antimicrobial properties and is studied in Ayurveda. Additionally, its oil is gaining attention in eco-friendly engineering applications, particularly as a potential bio-based cutting fluid in sustainable machining practices (Mradu, 2016).

The leaves of Putranjiva tree are shiny, dark green, simple, and roughly textured with an oblong shape. The fruits are round, stone-like drupes that usually contain a single seed. These seeds contain important compounds like glycosides such as glucojiaputin, glucocochlearin, glucoputranjivin, and glucocleomin which release a mustard-like smell due to the presence of isothiocyanate oil (Garg and Mitra 1968; Sengupta and Mukherjee 1968). Traditionally, both the leaves and fruits of the plant have been used in Thai medicine to relieve muscle sprains and joint pain. The entire plant has also been applied in treatments for general illness and bleeding disorders. Researchers have extracted several compounds from this plant, including two types of triterpenoids and four additional ones from its bark (Garg and Mitra, 1971). Additionally, a triterpene acid and a biflavonoid were obtained from an alcohol-based extract of its leaves. Because of these medicinal properties and bioactive compounds, Putranjiva roxburghii is considered a valuable plant for both traditional healing and scientific research. Figure 3.1 shows the Putranjiva roxburghii tree located on the NIT Silchar campus. This tree was observed and documented as part of the study for its relevance to the research. The leaves have cooling and fertility-enhancing properties, making them helpful in managing fever, congestion, and reproductive issues.

The seeds have a refreshing and slightly sour taste. The plant is known to support eye health, act as a vomit-inducer when needed, reduce inflammation, increase urine flow, and boost sexual desire. Traditionally, it has been used to treat various health problems, including mouth and stomach sores, skin swelling from heat, smallpox, burning sensations, eye disorders, excessive vomiting, elephantiasis, weakness, urinary problems, low sperm count, recurrent miscarriage, and infertility (Pandey and Fulara, 2022). From an engineering perspective, the oil extracted from its seeds is of particular interest due to its high content of fatty acids (notably 9-octadecenoic acid and ethyl oleate), which impart favorable lubrication characteristics. Previous studies have demonstrated the oil's potential as an eco-friendly industrial lubricant and its ability to enhance tribological performance, especially when chemically modified or blended with nanoparticles such as SiO_2. These attributes make Putranjiva roxburghii seed oil a promising candidate for sustainable machining applications, where improved lubricity, reduced wear, and environmental compatibility are required.

FIGURE 3.1 Putranjiva roxburghii tree at the NIT Silchar campus

The oil extracted from Putranjiva roxburghii seeds is rich in bioactive compounds, with 9-octadecenoic acid and ethyl oleate identified as major constituents. The oil exhibits a wide range of biological activities, including antibacterial, antioxidant, antiandrogenic, antidiabetic, antiarthritic, and anti-inflammatory effects (Emasushan and Britto 2018). The oil is also used in traditional medicine for treating a variety of ailments, including pain, rheumatism, and reproductive disorders. Additionally, the oil can be blended with diesel as an alternative fuel, offering similar efficiency and improved emissions (Mradu 2016). Epoxidized Putranjiva roxburghii oil, when blended with silicon dioxide nanoparticles, has been shown to enhance lubricity and reduce wear and friction in tribological applications. This suggests that Putranjiva roxburghii oil, especially in its epoxidized and nanoparticle-blended form, is a promising candidate for use as an industrial lubricant, offering a renewable and effective alternative to conventional synthetic oils (Singh et al. 2021).

3.3.2 Putranjiva roxburghii oil extraction process

The seeds of the Putranjiva roxburghii tree were collected directly from the campus of the National Institute of Technology (NIT) Silchar, located in Assam, India, at a geographical position of 24.50°N latitude and 92.510°E longitude, with an elevation of 114.68 meters above mean sea level. This region naturally supports the growth of Putranjiva roxburghii trees, providing an abundant and accessible source of seeds. As these trees are commonly found on the institute's grounds, the seed collection process was convenient, cost-free, and sustainable. After harvesting, the seeds were carefully cleaned to remove dust, soil, dry leaves, and other impurities to ensure cleanliness and maintain the quality of the raw material. Clean seeds are essential for obtaining pure and uncontaminated oil during the extraction process.

Following cleaning, the seeds were manually deshelled using a hammer to break the hard outer layer and access the inner kernel, which is the oil-bearing part of the seed. This process required precision to avoid damaging the kernels. Once the kernels were separated, they were laid out under sunlight and dried for approximately eight hours. This drying step was crucial to reduce the moisture content, which not only aids in improving the efficiency of oil extraction but also helps in preventing microbial growth and oil spoilage during storage. After adequate drying, the kernels were processed in a mechanical oil extraction unit, which applied pressure to extract the raw oil without involving chemical solvents. This method ensured that the natural properties of the oil were preserved, making it suitable for eco-friendly applications such as lubrication and cooling in machining processes. The figure presents a visual overview of the different stages of Putranjiva roxburghii seeds prior to oil extraction. Figure 3.2 (a) shows the tree bearing clusters of premature green fruits, representing the initial stage of seed development while still attached to the branches. Figure 3.2 (b) displays the mature seeds after harvesting, where the outer fruit layer has dried and been removed, leaving clean, hard seeds prepared for processing. Figure 3.2 (c) illustrates the embryo seeds just before oil extraction. Together, these images represent the complete pre-extraction preparation process of Putranjiva roxburghii seeds.

FIGURE 3.2 Putranjiva Roxburghii: (a) Premature seeds, (b) Mature seeds, (c) Embryo seeds

Figure 3.3 illustrates the systematic process involved in the extraction of Putranjiva roxburghii oil from its seeds. The process begins with the collection of seeds, where mature and healthy seeds are gathered for further processing. Following this, the seeds undergo a cleaning step to remove dirt, debris, and any foreign materials, ensuring purity and quality. The next stage involves seed conditioning, where the seeds are treated to achieve optimal moisture content and texture, which facilitates efficient oil extraction. The conditioned seeds are then subjected to crushing, where mechanical force is applied to break them down and release the oil content. Finally, the extracted oil is collected and stored, as depicted in the image showing a sample of the final oil product. This step-by-step method ensures maximum yield and quality of the bio-based oil used in various applications, including eco-friendly machining lubricants.

FIGURE 3.3 Flow diagram of oil extraction process

3.3.3 Preparation of nanofluids

The preparation of nanofluids begins with the extraction of base oil from Putranjiva roxburghii seeds using a standard mechanical extraction process. To enhance the oil's thermal and tribological properties, silicon dioxide (SiO_2) and graphene nanoparticles are incorporated as additives as shown in Figure 3.4. These nanoparticles are separately dispersed into the base oil at predetermined weight fractions of 0.25%, 0.5%, 0.75%, and 1.0%, resulting in the formulation of distinct nanofluid samples for further analysis. Silicon dioxide (SiO_2) and graphene nanoparticles were used to prepare vegetable-based nanofluids. These 99.9% purity nanoparticles were procured from Nano Research Labs, Jamshedpur. The physical properties of the nanoparticles as provided by the supplier are presented in Table 3.1. In each 20 gram of the vegetable (base) oil different wt.% (0.25%, 0.50%, 0.75%, and 1%) of SiO_2 and graphene nanoparticles were mixed separately. Also, hybrid nanofluids having equal proportions of both SiO_2 and graphene (0.125% SiO_2 + 0.125 % graphene, 0.25% SiO_2 + 0.25% graphene, 0.375% SiO_2 + 0.375% graphene, 0.5% SiO_2 + 0.5% graphene) nanoparticles were prepared. The nanoparticles were weighed using the weighing machine (Make: Sartonus, Model: BSA2245-CW) having a precision of 0.001g. Based on the weight of nanoparticles and Putranjival roxburghii oil (or any base oil), the % concentration of nanoparticles, C_{np} can determined by Singh et al. (2021).

$$C_{np} = \left\{ \frac{\frac{w_{np}}{\rho_{np}}}{\frac{w_{np}}{\rho_{np}} + \frac{w_{po}}{\rho_{po}}} \right\} \times 100, \tag{3.1}$$

Where w_{np} and ρ_{np} are the weight and density of nanoparticles, respectively, while w_{po} and ρ_{po} represent the weight and density of Putranjiva roxburghii oil, respectively.

To ensure that the nanoparticles are evenly spread out in the base fluid, a magnetic stirrer (Make: REMI, Model: 2-MLH) and a probe sonicator (Make: SONICS Vibra cell, Model: VCX750) were used. A magnetic stirrer comprises a rotating magnet and ensures proper mixing of the solution. At first, the base oil was put in a beaker. SiO_2 and graphene nanoparticles of different weight percentages were mixed individually. After that, a magnetic stirrer bar was put into the beaker and spun for 60 minutes at 900 revolutions per minute (RPM).

TABLE 3.1 Characteristics of nanoparticles

Characteristics	SiO_2	Graphene
Average particle size	30–50 nm	3–6 nm
Molar weight	60.08 g/mol	12.01 g/mol
Colour	White	Black
Bulk density	0.02–0.1 g/cm^3	2.2 g/cm^3
Specific surface area	650 m^2/g	120–150 m^2/g

In the next step, the prepared samples went through ultrasonication. The sonicator properly dispersed the nanoparticles in the liquid medium and broke the particles down to nano size. The probe of the sonicator that transmits energy in the form of ultrasonic waves was inserted into the sample; this helped in the dispersion of nanoparticles. A similar preparation method was followed by Chang et al. (2020) for the preparation of boron nitride nanotubes in solvents. The ultrasonication was carried out for 60 minutes. The two-step method is a common and affordable way to prepare nanofluids. First, nanoparticles are made separately using techniques like sol-gel or chemical precipitation. These nanoparticles are then mixed into a base fluid such as water, oil, or ethylene glycol. To ensure a proper mix and prevent the particles from clumping together, tools like magnetic stirrers, ultrasonication devices, or surfactants are used. In the first step, nanoparticles like SiO_2 or carbon nanotubes are prepared using chemical or physical methods. The complete process for the preparation of nanofluids is illustrated in Figure 3.5.

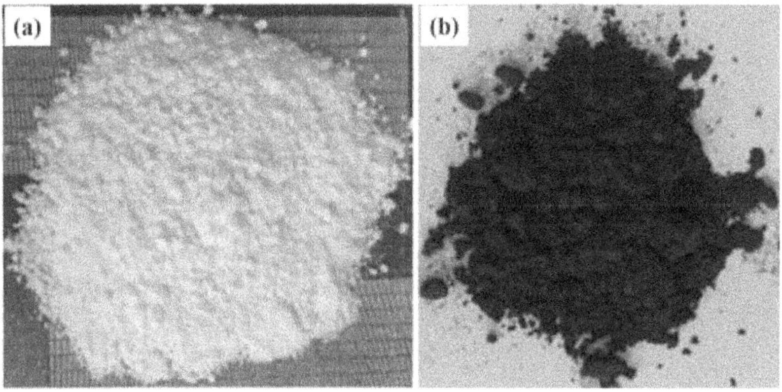

FIGURE 3.4 Nanoparticles: (a) SiO_2 and (b) Graphene

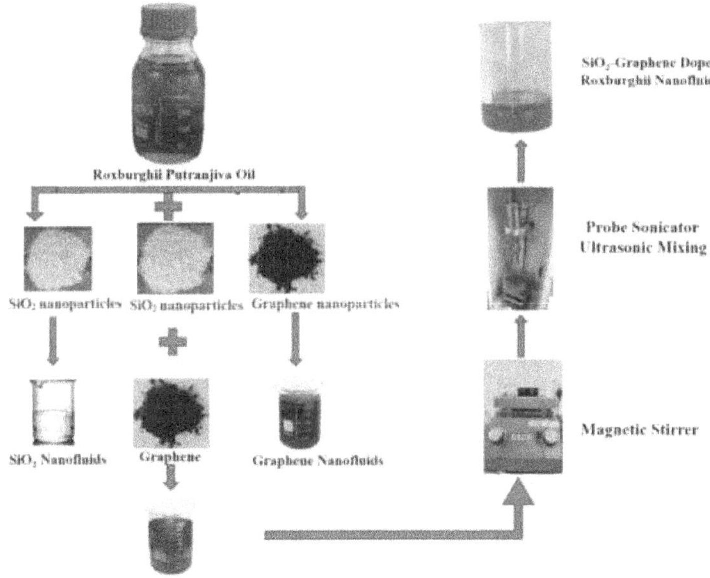

FIGURE 3.5 Preparation methodology of nano and hybrid nanofluids

In the second step, these particles are added to the base fluid. Stirring and ultrasonication help in mixing, and surfactants may be added to keep the mixture stable. One challenge in this method is that nanoparticles can settle over time. To prevent this, steps like adjusting pH, adding surfactants, or carefully controlling sonication time are taken. These measures help maintain the fluid's stability and improve properties like thermal conductivity and viscosity.

3.3.4 Preparations of emulsions

An emulsion is a mix of two liquids that do not normally combine, where one liquid is spread out as small droplets in the other with the help of stirring and an emulsifier. There are two types of emulsion: (1) Oil-in-water, where oil is dispersed in water and (2) Water-in-oil, where water is dispersed in oil.

In this study, oil-in-water emulsions were prepared using six different ratios of water to oil: 100:05, 100:10, 100:15, 100:20, 100:25, and 100:30 ml. Here, 100 ml of distilled water was used in each case as the base, and different volumes of vegetable oil were added. Lecithin, an emulsifying agent, was added at 1–2% of the total volume to help mix the oil and water and keep the emulsion stable. The emulsions were prepared using a magnetic stirrer. Water was poured into a beaker with a stir bar, and the oil mixed with emulsifier was slowly added while stirring at 1000 RPM. Mixing continued for about 40–45 minutes until a uniform emulsion was formed. This process was repeated for all six formulations. All emulsions were stored in sealed containers, labelled by their water to-oil ratio, and later tested for properties like stability, viscosity, and thermal properties. Figure 3.6 shows the emulsion-making process and prepared samples.

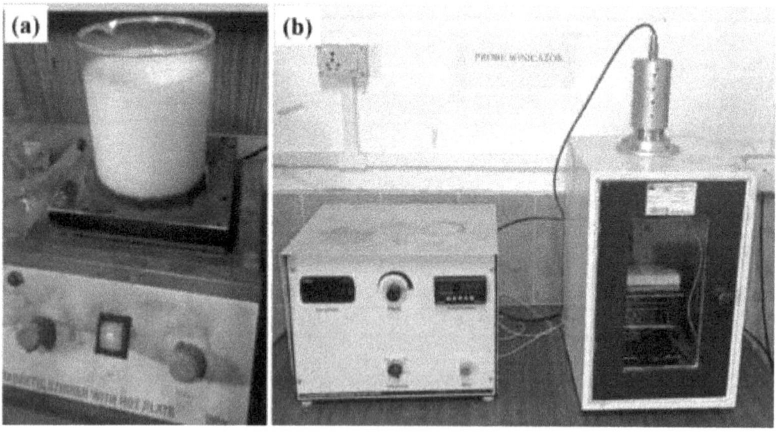

FIGURE 3.6 Preparation of emulsion: (a) Magnetic stirrer and (b) Probe Sonicator

3.3.5 Preparations of nanoemulsion

Figure 3.7 illustrates the preparation process of nanoemulsion cutting fluids using Putranjiva roxburghii oil, distilled water, an emulsifying agent, and nanoparticles. Initially, the oil and water are mixed with an emulsifier using a magnetic stirrer to form stable emulsion cutting fluids of various ratios. Nanoparticles are then added to these emulsions, and the mixture is subjected to probe sonication to ensure uniform dispersion.

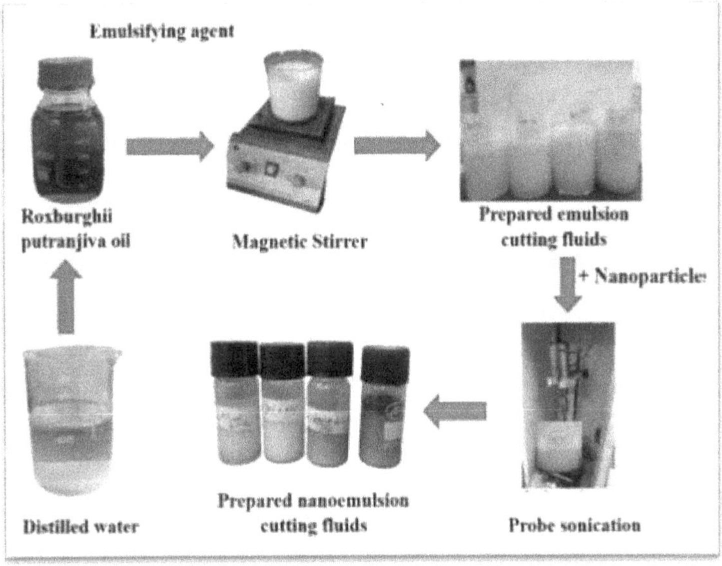

FIGURE 3.7 Flow diagram of preparation of nanoemulsion

Through this method, nanoemulsion-based cutting fluids are generated, which are used in eco-friendly machining applications due to their enhanced lubrication and heat dissipation properties. The methodology for the preparation of nanoemulsion cutting fluids is illustrated in the figure.

3.4 Characterization Methods Employed for Lubricants

This section provides a detailed discussion of the equipment and methodologies employed in the preparation and characterization of samples used in the present research. It includes a comprehensive overview of the various instruments utilized for sample synthesis, such as the oil extraction unit, magnetic stirrer, and probe sonicator, as well as analytical tools like the FTIR spectrometer. Additionally, the methodologies followed for preparing nanofluids and ensuring uniform dispersion of nanoparticles are described in detail. The characterization techniques adopted to examine the material's physical behavior and chemical composition of the prepared samples are also discussed to establish the reliability and effectiveness of the materials developed for environmentally friendly machining applications.

3.4.1 Physical properties analyzer

The viscosity of nanofluids prepared by dispersing SiO_2 and graphene nanoparticles in Putranjiva roxburghii oil was evaluated using a rheometer (Make: Anton Paar, Model: MCR 101). The viscosity measurements were carried out over a temperature range of 30 °C to 120 °C, which reflects typical machining conditions. Shear rate scans were performed from 1 to 1000 s^{-1}, with 40 data points recorded at 6-second intervals. The resulting torque values ranged from 2.5 μNm to 2600 μNm across the shear rate range, maintaining an accuracy of ±1%. The rheometer used a concentric cylinder geometry (SS type, CC27-SN40306), suitable for analyzing low-viscosity fluids, with a sample volume of 7 ml for each test. In addition to viscosity, the density of Putranjiva roxburghii oil was measured at 20 °C and 40 °C using a density meter (Make: ViscoDens, Model: VDN 300, USA). The pour point and cloud point were assessed following ASTM D97 standards using the HAMCO pour and cloud point apparatus (Model: WBN 300). Furthermore, the flash and fire points were determined with a Pansy-Martens closed cup tester.

3.4.2 Thermal properties analyzer

The thermal analyzer was used to measure the thermal conductivity and volumetric heat capacity. The thermal analyzer equipment was based on the transient hot wire approach. The sensor present in the analyzer was in the form of a needle. The needle was 1.27 mm in diameter, 60 mm long and made of stainless steel. The error in the measurement could go up to ±5%. Five repetitions were done for measuring thermal conductivity and volumetric heat capacity. To reduce the measurement error caused by free convection effects at the measurement sensor, a 5-minute interval was maintained between each measurement. The KD2 Pro device (Make: Decagon Devices Inc., Model: KD2 Pro, USA) was utilized to determine the heat-related characteristics of the developed emulsions and nanofluids, as shown in Figure 3.8. This device is widely used for precise determination of thermal conductivity, thermal diffusivity, and specific heat capacity of fluids and semi-solids.

Specific heat represents how much heat energy a fluid can hold without a significant rise in temperature. The addition of graphene (low specific heat) and SiO_2 (moderate specific heat) can slightly alter the specific heat of the base fluid. Although there may be a marginal reduction, the enhanced thermal conductivity typically compensates for this in machining environments where rapid heat dissipation is critical.

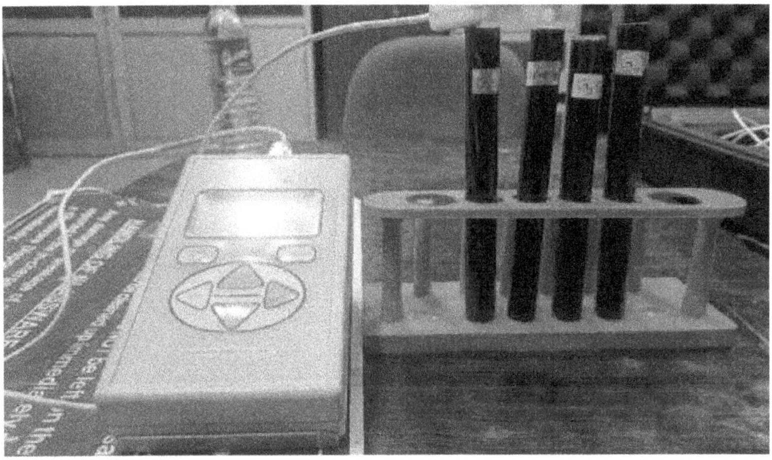

FIGURE 3.8 KD2 Pro thermal property analyzer

3.4.3 Wettability Study

The lubricating behavior of the developed fluids was examined through a contact angle test by determining the contact angle between a liquid droplet and a solid substrate. This contact angle indicates the extent of fluid spreading on the surface, with smaller angles representing enhanced wetting and therefore improved lubrication capability. The measurements were carried out using a Goniometer (Make: Holmarc, Model: HO-IAD-CAM-01B, India) as shown in Figure 3.9, an optical instrument specifically designed to capture and analyze the shape of liquid droplets on solid substrates.

In this study, the contact angle of each cutting fluid was assessed. A 2 μl droplet of the cutting fluid was carefully placed on a polished Inconel 718 superalloy substrate, which had a surface roughness (R_a) value of 0.0141 μm. The droplet was allowed to stabilize for 10 seconds before the image was captured and the contact angle was precisely measured using the goniometer. This procedure provided insights into the spreading behavior and surface interaction of the fluids under investigation.

FIGURE 3.9 Goniometer used for measuring contact angle

3.4.4 Rheological Study

A rheological study is the scientific investigation of how a material flows and deforms under applied forces. It focuses on understanding the viscosity, elasticity, and plasticity of fluids and soft solids, especially under different conditions such as temperature, shear rate, or pressure. In the context of vegetable oils or cutting fluids, a rheological study helps determine how the fluid behaves when it is subjected to mechanical forces during machining. For the rheological study of the nanofluids, a rheometer (Make: Anton Paar, Model: Physica MCR 101, Austria) was used, as shown in Figure 3.10. This instrument is capable of accurately measuring viscosity, shear stress, and flow behavior under varying shear rates, providing valuable insights into the fluid's deformation and flow characteristics.

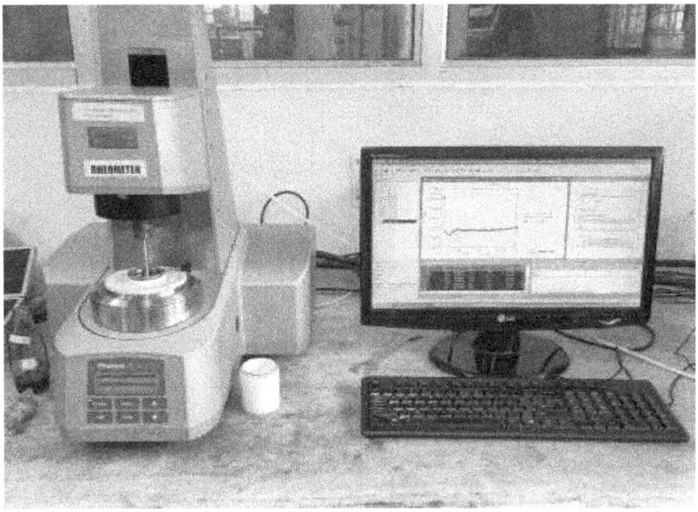

Figure 3.10 Rheometer setup

3.5 Experimental Setup for Testing the Performance of Lubrication

All machining experiments were performed using a conventional milling machine (Make: BFW Inder Industries, Model: HF-2, India) located at the Central Workshop of the Indian Institute of Technology (IIT) Guwahati, India. This machine has a rated spindle power of 5.5 kW and is suitable for a range of machining operations, including face milling. For the present study, the machine was equipped with additional attachments for air cooling and MQL systems to facilitate sustainable machining conditions. For air cooling condition, an air compressor was utilized that facilitated compressed air was continuously supplied by air flow regulator comprising a nozzle of diameter 3 mm to the machining area. The nozzle was positioned 20 mm away from the cutting tool's tip. Figure 3.11 illustrates the entire milling machine setup, highlighting the integration of both air and MQL delivery systems. Face milling operations were conducted on Inconel 718, a nickel alloy recognized for its poor machinability, under different lubrication and cooling setups to investigate aspects like tool heat generation, surface finish quality, and chip removal behavior. Under the MQL condition, vegetable oil-based cutting fluids were used. The cutting fluid was stored in a container placed at a vertical height of 1220 mm above the machining area, denoted as *h*.

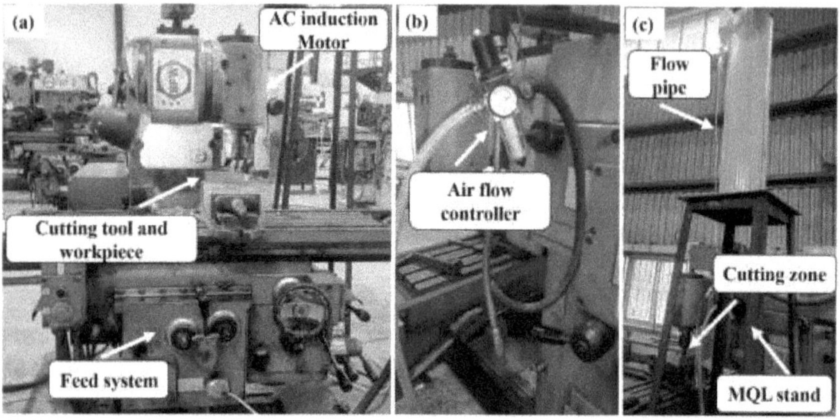

FIGURE 3.11 Experimental set up: (a) Vertical milling machine, (b) air cooled setup and (c) MQL setup

Additionally, the container was positioned 915 mm away horizontally (*l*) from the cutting point, creating an inclined path for fluid delivery. A plastic tube connected the fluid container to the machining zone, enabling the oil to travel under gravity. To precisely control the amount of fluid reaching the cutting zone, a flow regulator was installed in the path. The regulator was calibrated to allow 20 drops (approximately 1 *ml*) of cutting fluid, thereby ensuring minimal fluid usage in line with MQL principles. The schematic arrangement of the MQL system is depicted in Figure 3.12.

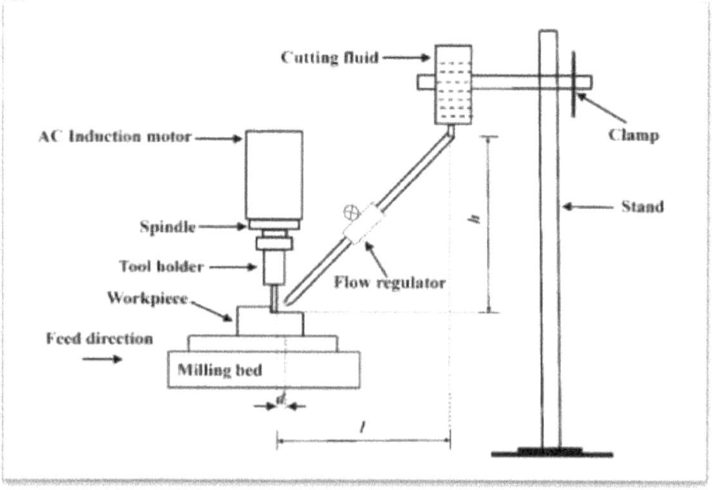

FIGURE 3.12 Schematic of the MQL setup

The fluid delivery nozzle was located only 5 mm away from the tip of the tool, guaranteeing effective fluid flow directly at the cutting tool-workpiece interface. This accurate positioning aided in lowering friction and heat generation during machining while also reducing fluid consumption.

3.5.1 Workpiece and tool materials

Inconel 718 was selected for this study because of its extensive application in critical sectors such as aerospace and power plants, attributed to its high strength, resistance to corrosion, and thermal stability. Nonetheless, machining this alloy is challenging due to its poor thermal conductivity, significant work hardening behavior, and propensity to increase tool wear while compromising surface quality. These challenges make it an ideal material for evaluating different cooling and lubrication strategies such as MQL, cryogenic cooling, compressed-air cooling, and bio-based cutting fluids. Its common use in research also makes it a suitable material for comparing machining techniques and tool performance.

By focusing on this material, the research aims to contribute valuable insights into environmentally friendly and efficient machining practices that are directly relevant to modern manufacturing industries (Pereira et al. 2024; Sebbe et al. 2025; Wu and Lin 2021). The workpiece selected for face milling was Inconel 718 superalloy. The Inconel 718 workpiece materials in the form of rectangular block was purchased from local market with dimensions of 125 mm × 100 mm × 13 mm. The elemental compositions of workpiece material as provided by the supplier were 52.82% nickel, 19% chromium, 17.8% iron, 5.43% niobium and tantalum, 3.04% molybdenum, 0.98% titanium, 0.5% aluminium, 0.23% cobalt, 0.08% manganese, 0.08% silicon and 0.04% carbon. Figure 3.13 displays the Inconel 718 plate used as the workpiece material in the present study.

The machining of Inconel 718 was carried out using a 20 mm diameter cemented carbide tool bit with Widia PA 145 inserts. These inserts had a TiAlN (Titanium Aluminium Nitride) coating on a carbide base. The insert featured an axial rake angle of 4°, a radial rake angle of 6°, and a clearance angle of 10°. These rake angles were determined using a coordinate measuring machine. (Make: Accurate Spectra, Model: Spectra), while the clearance angle was measured with a tool maker's microscope (Make: Sipcon, Model: SMT).

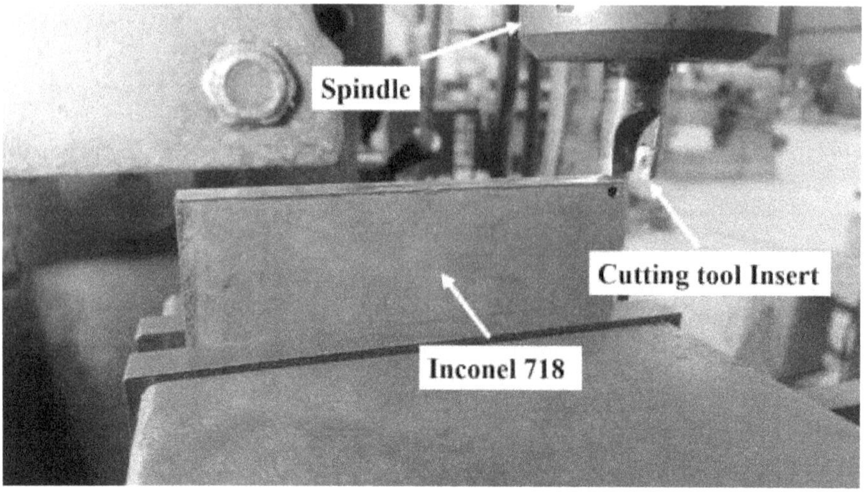

FIGURE 3.13 Inconel 718 plate as a workpiece material

TiAlN coatings are widely recommended for machining difficult-to-cut nickel based superalloys such as Inconel 718 due to their high hot hardness, excellent oxidation resistance, and superior thermal stability (Shalaby et al. 2015; Ezugwu and Bonney 2004). These properties enable the tool to withstand the extreme temperatures generated during machining and reduce tool wear mechanisms such as adhesion, abrasion, and diffusion. The coating also provides a low-friction interface, improving chip evacuation and enhancing surface finish. Several studies have demonstrated the effectiveness of TiAlN-coated tools in maintaining cutting-edge integrity and extending tool life when machining Inconel 718 (Pramanik et al. 2017). TiAlN coatings help protect the tool from heat and wear by forming a stable alumina (Al_2O_3) layer at high temperatures, which acts as a thermal barrier. This helps retain tool hardness, reduce friction, and prevent material sticking to the tool surface. As a result, tool life is extended, surface finish improves, and machining becomes more stable.

3.5.2 Temperature measurement

To evaluate the thermal response during the actual machining process, the temperature generated at the cutting zone was recorded using a thermal infrared imaging device (Make: Testo, Model: 868, Germany) as shown in Figure 3.14. This non-contact thermal camera offers a high thermal sensitivity of less than 0.08 °C (80 mK) and an accuracy of ±2 °C, making it suitable for detecting small temperature variations in dynamic cutting conditions. Temperature readings were captured at five distinct points during the movement of the machine table, and the average value was calculated to represent the overall thermal behavior under each test condition.

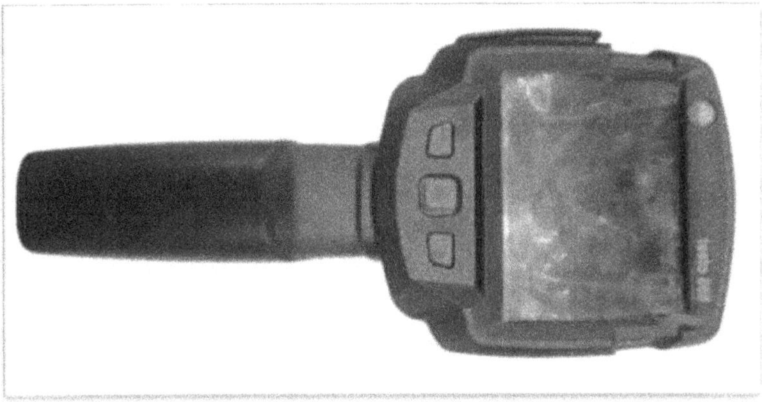

FIGURE 3.14 Thermal infrared imaging device

3.5.3 Surface roughness measurement

To evaluate the surface roughness of the machined Inconel 718 workpieces, both contact and non-contact measurement techniques as shown in Figure 3.15, were employed for accurate and comprehensive surface characterization. A contact-type surface roughness measuring instrument, Pocket Surf (Make: Mahr Federal, Model: PMD-90101, USA), was used to measure average surface roughness (R_a) by physically tracing the surface profile with a stylus. This device provides quick and portable measurements, making it suitable for on-site evaluations. In addition, a non-contact 3D optical profilometer (Make: Taylor Hobson, Model: Talysurf CCI Lite, United Kingdom) was utilized to capture detailed surface topography without physically touching the surface. This advanced instrument uses white light interferometry to generate high-resolution three-dimensional surface maps, enabling the analysis of parameters such as peak height, valley depth, and texture. The combination of both techniques ensured accurate, reliable, and in-depth assessment of surface quality under different machining conditions.

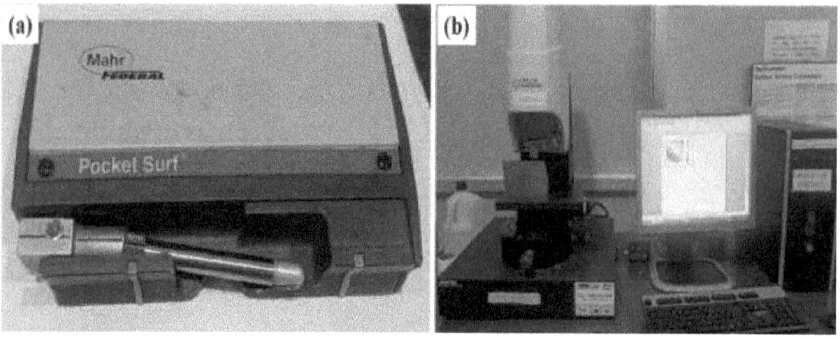

FIGURE 3.15 Surface roughness measuring instrument: (a) contact-type pocket surf (b) non-contact 3D optical profilometer

3.5.4 Zoom microscope

The chips produced during the machining process were analyzed using a Stereo Microscope (Make: Nikon, Model: SMZ25, Japan), a high-performance optical instrument specifically designed for three-dimensional observation of samples at low to medium magnification as shown in Figure 3.16. The Nikon SMZ25 is a cutting-edge stereo zoom microscope that offers a high zoom ratio (25:1) and excellent resolution, enabling clear visualization of surface features, textures, and chip morphology in great detail.

FIGURE 3.16 Zoom microscope

3.5.5 Tool angle measurement

The cutting insert used in this study was designed with a precise geometric configuration consisting of a 4° axial rake angle, a 6° radial rake angle, and a 10° clearance angle. To ensure accurate measurement of these angles, advanced metrology equipment was employed. The axial and radial rake angles were measured using a Coordinate Measuring Machine (CMM) (Make: Accurate Spectra, Model: Spectra, India), a high-precision device that utilizes a probing system to capture the exact spatial coordinates of complex geometries as shown in Figure 3.17 (a). The CMM provides three-dimensional measurements with micron-level accuracy, making it ideal for evaluating critical tool angles and dimensions.

The clearance angle was measured using a Tool Makers Microscope (Make: Sipcon, Model: SMT, India), a specialized optical instrument designed for precise angle and edge measurement of small machined components. This microscope offers a clear magnified view of the cutting edge and allows the user to measure angles accurately by rotating a calibrated stage, making it particularly useful for analyzing the flank surface of cutting tools. Figure 3.17 (b) shows the actual image Tool Makers Microscope device.

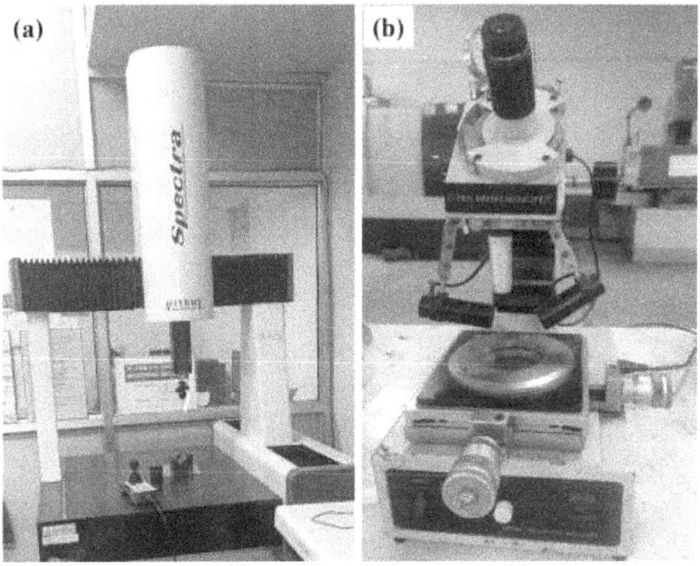

FIGURE 3.17 Tool angle measuring instruments: (a) Coordinate Measuring Machine, (b) Tool Makers Microscope

3.5.6 Measurement of axial Force

A 20 HP friction stir welding machine (Make: Royal Machine Tool, Model: RMV-1800, India) was adopted for performing the milling operations. The FSW machine shown in Figure 3.18, utilized for measuring the axial force during the milling operations. The axial force generated during the operation was measured using force and torque recording sensor (Make: Digitech, Model: MP31C09). To monitor the cutting forces during the milling operation, a dynamometer was integrated into the setup. This precision instrument is capable of measuring real-time machining forces, including axial force (thrust force) and torque, which are critical parameters for understanding tool performance, material behavior, and process efficiency. The dynamometer was mounted beneath the workpiece, enabling the direct capture of force data during the interaction between the cutting tool and the Inconel 718 plate.

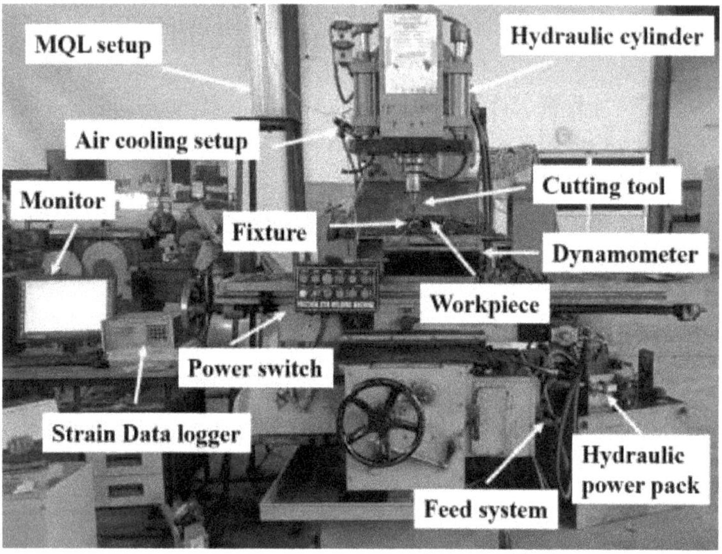

FIGURE 3.18 FSW machine with dynamometer setup

A suitable fixture comprising two angle plate was designed and mounted in the machine to hold the rectangular block shaped Inconel 718 workpiece. Machining was conducted in three different conditions, viz., dry, air, air cooling and nanoemulsion based MQL cooling. The machining variables were spindle speed (N), feed rate (f_r), and depth of cut (d_c). Milling was conducted at three spindle rotations, viz., 233, 360 and 462 revolutions per minute (RPM). Two feed rates were employed, viz., 13 and 20 mm/min. Depth of cut values used were 0.1 mm and 0.2 mm. For air cooling conditions, an air compressor was utilized that facilitated compressed air supply at different pressure variation viz. 0.5, 1.0 and 1.5 bar. During the experiments, compressed air was continuously delivered to the machining zone through a flow regulator equipped with a 3-mm diameter nozzle. The distance between the nozzle and the cutting tool tip was 20 mm. In MQL conditions, the machining was done using base emulsion (100 ml water + 20 ml Putranjiva roxburghii oil) and three distinct nanoemulsions based cutting fluids.

3.6 Optimization Method Employed in the Study

Optimization techniques are methods used to find the best possible solution to a problem, usually by maximizing or minimizing a certain objective, such as cost, efficiency, time, or quality. These techniques help in making better decisions by adjusting the input variables to get the most favourable outcome. Optimization is used in many fields like engineering, business, computer science, and manufacturing. In this experimental work an evolutionary technique used namely Box's evolutionary optimizations methods was used.

To determine the optimal combination of machining input parameters, the evolutionary optimization method proposed by G.E.P. Box in 1957 was employed. It is a method used for finding the optimal solution in multi-dimensional design spaces (Deb 2012). This technique uses a total of 2^N+1 evaluation points for N decision variables. 2^N number of points are located at the corners of an n-dimensional hypercube cantered around a base point. This technique operates by evaluating function values at 2^N+1 points. Starting with an initially assumed optimal point x_0, a hypercube is generated around this point with 2^N corner points. The function values at these 2^N+1 points are computed, and the best-performing point is selected as the new x_0. In subsequent iterations, a new N-dimensional hypercube is formed around the updated x_0, and the process is repeated by evaluating the function values at the new corners and identifying the most optimal point. This continues until two successive iterations yield the same best point, after which the hypercube size is reduced, and the cycle repeats. The iterations proceed until the hypercube becomes sufficiently small, indicating convergence.

In this method, table feed and spindle speed are chosen as the two main machining parameters for optimization. Table feed values are labeled as x_1, x_2, and x_3 (in mm/min), while spindle speeds are marked as y_1, y_2, y_3, and y_4 (in RPM). The process starts from a central point, which uses the middle values of both parameters: x_2 for feed and y_3 for speed. This starting point is noted as (x_2, y_3).

Next, a rectangular area (a 2D box) is formed around this central point. The four corners of this box are (x_1, y_4), (x_3, y_4), (x_1, y_2), and (x_3, y_2), as shown in Figure 3.19. Surface roughness is measured at each of these four points, plus the center, making five total evaluations. After comparing the results, the point (x_1, y_2) gives the lowest surface roughness. So, this point is selected as the new best point. A smaller box is then created around this new point (x_1, y_2), shown by dashed lines in the figure. The same evaluation process is repeated for the new box checking all corners and the center. The point with the best result becomes the new starting point for the next round. This process continues, with the box getting smaller each time, until no further improvement is found or the box becomes very small. The final best point gives the optimal combination of feed rate and spindle speed.

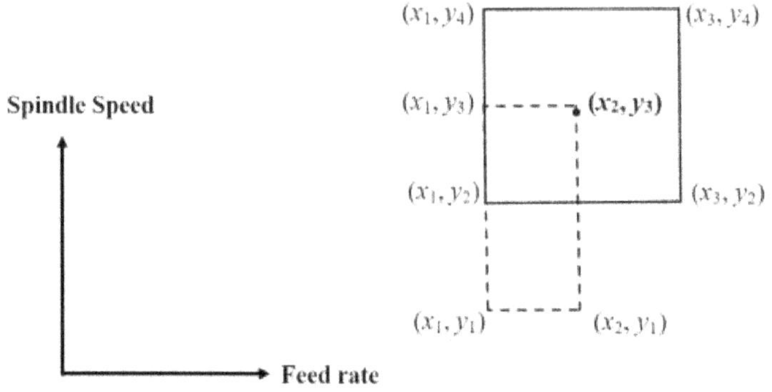

FIGURE 3.19 Design of experiments as per Box's evolutionary optimization

3.7 Characterization Techniques for Materials

Characterization techniques refer to a set of scientific methods and tools used to analyse the physical, chemical, thermal, and structural properties of materials. In the context of nanofluids and emulsions, these techniques help researchers understand how nanoparticles interact with base fluids, how stable the formulations are, and how they behave under different conditions.

The primary purpose of characterization is to validate the successful preparation of nanofluids and to evaluate their performance-relevant properties. These techniques provide vital data such as viscosity, thermal conductivity, particle dispersion, and chemical composition, which are essential for assessing the suitability of nanofluids in engineering applications like machining, lubrication, and heat transfer.

3.7.1 X-ray diffraction (XRD) analysis

The X-ray diffraction (XRD) analysis was performed using a fully computerized PANalytical powder X-ray diffractometer (Make: PANalytical, Model: PW3040/60 MPD, Netherlands), as shown in Figure 3.20. In this work, XRD has been employed for determine the crystallographic structure, phase composition, and average crystallite size of SiO_2 and graphene nanoparticles. It helps identify whether the nanoparticles are crystalline or amorphous and reveals specific phases present in the sample. Using Scherrer's equation, the average particle size of nanoparticles can be evaluated by (Bahari et al., 2022).

$$D_{np} = \frac{0.9\lambda}{\beta \cos\theta}, \quad (3.2)$$

where D_{np} is the size of the nanoparticle, λ is the wavelength, β is the total peak broadening and θ is the diffraction. The analysis utilized an XRD, specifically a PAN Analytical Powder X-ray diffractometer with CuKα radiation ($\lambda = 1.5406$ Å). The diffraction data were collected over a 2θ range of 5° to 90°, with an incremental step of 0.02°. The *d*-spacing values of the atomic layers in the developed materials were computed using Bragg's Law as,

$$n\lambda = 2d\sin\theta. \quad (3.3)$$

Here, n represents the diffraction order, λ is the wavelength of the incident X-rays (1.5406 Å), d is the interplanar distance, and θ is the angle of incidence. For this analysis, an X-ray beam with $\lambda = 1.5406$ Å was directed at the composite and nanocomposite specimens.

3.7.2 Field Emission Scanning Electron Microscopy (FESEM)

Field Emission Scanning Electron Microscopy (FESEM) utilizes a finely focused electron beam to obtain high-resolution images of a material's surface features. Figure 3.21 illustrates the FESEM configuration. In this research, FESEM was employed to examine the surface morphology of nanoparticles and tool wear under various cutting conditions. Additionally, elemental composition analysis was performed using the integrated Energy Dispersive X-ray (EDX) detector installed on the Carl Zeiss Sigma 300 microscope.

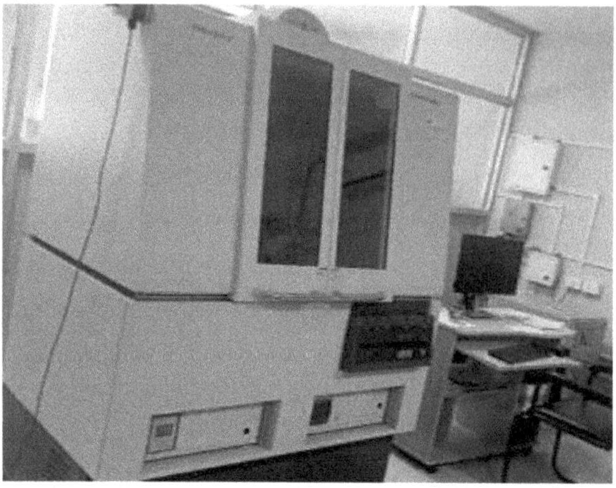

FIGURE 3.20 PAN analytical X'Pert Pro XRD machine

A major advantage of SEM lies in its exceptionally large depth of field, allowing clear images at magnifications ranging from 10X to 300,000X. The surface structures of the cutting tools were investigated using field emission scanning electron microscopy. (Make: Zeiss, Model: Sigma 300, Germany)

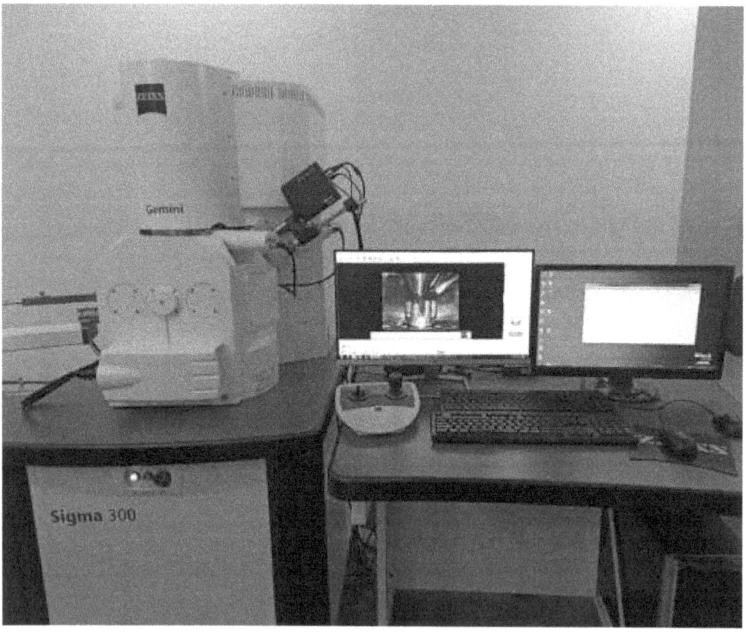

FIGURE 3.21 FESEM machine Sigma 300

3.7.3 Fourier Transform Infrared Spectroscopy (FTIR)

This method is employed to analyze the infrared absorption or emission spectra of solids, liquids, and gases by acquiring high-resolution spectral data across a broad frequency range. In crystalline materials, atoms or ions are arranged in a structured lattice and are bonded through chemical interactions, functioning similarly to mechanical springs.

The FTIR of various samples was carried out using Thermo Scientific FTIR Spectrometer (Make: Thermo Scientific, Model: Nicolet iS5, USA) as shown in Figure 3.22. During analysis, infrared radiation interacts with the sample, causing molecular bonds to vibrate at specific frequencies, which are determined by the types of atoms involved and the nature of the chemical bonds. These characteristic absorption frequencies indicate the excitation of vibrational modes within the molecules and are unique to particular bond types and atomic groupings. The technique also involves photon-induced transitions between vibrational energy levels of molecules and solids within the infrared region of the electromagnetic spectrum. This region typically spans from 4000 to 400 cm^{-1}, which conveniently corresponds to the energy gaps found in many semiconductor materials, allowing them to be effectively characterized using this approach.

FIGURE 3.22 FTIR Spectrometer

3.7.4 Ultraviolet–visible (UV-Vis) Analysis

Ultraviolet–visible (UV-Vis) spectroscopy was employed to evaluate the suspension stability of nanoparticles dispersed in Putranjiva roxburghii oil. This technique uses monochromatic ultraviolet light to determine the absorbance of a substance within a liquid medium, where the absorbance value directly correlates with the concentration of the dispersed particles. Ensuring uniform dispersion of nanoparticles is essential for formulating stable and effective nanofluids. In this study, UV absorbance was measured for various concentrations of graphene and SiO_2 nanoparticles suspended in Putranjiva roxburghii oil. The analysis was performed using a UV-Vis spectrophotometer (Make: Varioskan LUX, Model: LAT module, USA) shown in Figure 3.23. Diluted nanofluid samples were placed in cuvettes small, transparent containers specifically designed for holding liquid samples during spectroscopic measurements. This setup enabled precise evaluation of the optical absorption characteristics of graphene and SiO_2 nanoparticles. The spectral scanning range of the instrument was configured from 200 to 900 nm to capture relevant absorbance data.

The fundamental principle involves the dispersion of electromagnetic waves from a source, which then pass through the sample. A portion of the transmitted radiation is subsequently measured and correlated with the incident wavelength using an appropriate detection system, allowing for the characterization of the material's optical and electronic properties.

FIGURE 3.23 UV-Vis spectrophotometer

3.8 Summary

This chapter provided an in-depth description of the materials, machinery, and experimental methods utilized to study the eco-friendly milling of Inconel 718 using dry, air-cooled, and bio-based MQL approaches. Inconel 718 was chosen as the workpiece material due to its significance in aerospace and high-temperature environments, where improving machinability alongside sustainability is essential. A detailed description of the cutting tools, machining parameters, and cooling/lubrication strategies was provided to establish the experimental scope. Using bio-based nano-lubricants in MQL supports the aim of promoting green manufacturing. Accurate instruments were used to collect reliable data. The methods used here provide a strong base for the experiments. They are expected to give useful results about how eco-friendly machining affects the quality and performance of Inconel 718 milling.

References

Bahari, A., Sadeghi-Nik, A., Shaikh, F.U.A., Sadeghi-Nik, A., Cerro-Prada, E., Mirshafiei, E. and Roodbari, M., 2022. Experimental studies on rheological, mechanical, and microstructure properties of self-compacting concrete containing perovskite nanomaterial. *Structural Concrete*, *23*(1), pp.564-578.

Balkrishna, A., Nain, P., Joshi, M., Khandrika, L. and Varshney, A., 2021. Supercritical fluid extract of putranjiva roxburghii wall. Seeds mitigates fertility impairment in a zebrafish model. Molecules, 26(4), p.1020.

Chang, M.S., Jang, M.S., Yang, S., Yu, J., Kim, T., Kim, S., Jeong, H., Park, C.R. and Jeong, J.W., 2020. Electrostabilized homogeneous dispersion of boron nitride nanotubes in wide-range of solvents achieved by surface polarity modulation through pyridine attachment. Nano Research, 13, pp.344-352.

Deb K. Optimization for Engineering Design: Algorithms and Examples. PHI Learning Private, Limited, New Delhi, India 2012.

Emasushan, M. and Britto, J.S., 2018. Seed oil extraction and phytochemical analysis of Putranjiva roxburghiiwall (Putranjivaceae). Journal of Emerging Technologies and Innovative Research, 5(10).

Ezugwu, E.O. and Bonney, J., 2004. Effect of high-pressure coolant supply when machining nickel-base, Inconel 718, alloy with coated carbide tools. *Journal of materials processing technology*, *153*, pp.1045-1050.

Garg, H.S. and Mitra, C.R., 1968. Putranjiva roxburghii wall II: Triterpenes of the trunk bark. *Phytochemistry*, *7*(11), pp.2053-2055.

Garg, H.S. and Mitra, C.R., 1971. Roxburghonic acid—A friedelane triterpenoid keto acid of the leaf of Putranjiva roxburghii. *Phytochemistry*, *10*(4), pp.865-869.

Kacher J, Landon C, Adams BL, Fullwood D. Bragg's Law diffraction simulations for electron backscatter diffraction analysis, Ultramicroscopy. 2009; 109: 1148–1156.

Kumar, N., 2020. Phytochemistry and medicinal value of putranjiva roxburghii wall. Advances in pharmaceutical biotechnology: recent progress and future applications, pp.133-144.

Lv, T., Huang, S., Hu, X., Ma, Y. and Xu, X., 2018. Tribological and machining characteristics of a minimum quantity lubrication (MQL) technology using GO/SiO 2 hybrid nanoparticle water-based lubricants as cutting fluids. *The International Journal of Advanced Manufacturing Technology*, *96*, pp.2931-2942.

Mradu, G., 2016. A review of pharmacological properties, pharmacognosy and therapeutic actions of Putranjiva roxburghii Wall (P. roxburghii). Int J Herb Med, 4(6), pp.104-8.

Pandey, R. and Fulara, B.T., 2022. The therapeutic uses of Putranjiva roxburghii. Int J Vet Sci Anim Husb, 7(4), pp.51-54.

Pereira, R.B.D., Pérez-Ruiz, J.D., Gómez-Escudero, G., Calleja-Ochoa, A., Villarrazo-Rubia, N., Brandão, L.C. and Lacalle, L.N.L.D., 2024. Helical milling hole finishing of Inconel 718 parts manufactured by laser powder bed fusion. *Progress in Additive Manufacturing*, pp.1-17.

Ramírez Fernández, M.P., Mazón, P., Gehrke, S.A., Calvo-Guirado, J.L. and De Aza, P.N., 2017. Comparison of two xenograft materials used in sinus lift procedures: Material characterization and in vivo behavior. *Materials*, *10*(6), p.623.

Sarath, P. and Sudha, B.R., 2019. A comparative evaluation of phytochemicals in bark, leaves and seeds of Putranjiva roxburghii Wall. (Putranjivaceae). J Pharm Phytochem, 8(1), pp.1162-6.

Sebbe, N.P.V., Silva, F.J., Iglesias, I., Jesus, A. and Sales-Contini, R.C.M., 2025. Experimental Study on the Performance in Milling Operations of Inconel® 718 With TiN/TiAlN-Coated Tools. *Journal of Tribology*, *147*(4).

Sengupta, P. and Mukherjee, J., 1968. Terpenoids and related compounds—XI: The structure of roxburgholone, a new triterpenoid constituent of Putranjiva roxburghi. *Tetrahedron*, *24*(20), pp.6259-6264.

Singh, R., Sah, N.K. and Sharma, V., 2021. Development and characterization of unitary and hybrid Al2O3 and ZrO dispersed Jatropha oil-based nanofluid for cleaner production. Journal of Cleaner Production, 317, p.128365.

Singh, Y., Singh, N.K., Sharma, A. and Badhotiya, G.K., 2022. Tribo-mechanism analysis of epoxidized Putranjiva oil (Putranjiva roxburghii) and understanding lubricity with silicon oxide nanoparticle addition. *Proceedings of the Institution of Mechanical Engineers, Part J: Journal of Engineering Tribology*, *236*(4), pp.648-658.

Szkliniarz, A., Moskal, G., Szkliniarz, W. and Swadźba, R., 2015. Improvement of oxidation resistance of Ti–47Al–2W–0.5 Si alloy modified by aluminizing method. *Surface and Coatings Technology*, *277*, pp.270-277.

Wu, T.Y. and Lin, C.C., 2021. Optimization of machining parameters in milling process of inconel 718 under surface roughness constraints. *Applied Sciences*, *11*(5), p.2137.

CHAPTER

4

Characterizations of Putranjiva Roxburghii Oil with Addition of SiO_2 and Graphene Nanoparticles

4.1 Introduction

In machining processes, cutting fluids serve to decrease friction and dissipate heat at the interface between the tool and the workpiece. Historically, petroleum-based fluids have been commonly employed for this purpose, but recently growing health concerns are forcing the researchers to replace them by vegetable oils. For the sake of sustainability, it is important to evaluate the performance, cost and environmental impact of a cutting fluid (Dixit et al. 2012; Gupta et al. 2023). Conventional cutting fluids include toxic chemicals and additives that may cause skin irritations, respiratory issues and long-term health problems. Vegetable-based cutting fluids derived from biodegradable, renewable resources such as soybeans, canola and sunflowers mitigate this problem. Various vegetable-based cutting fluids, including canola, sunflower, coconut, jatropha, and palm oils, have been investigated for their advantages in machining processes. Kuram et al. (2013) employed canola, sunflower, and a semi-synthetic oil as cutting fluids during the milling of AISI 304 stainless steel.

They found that canola oil was preferable for minimizing specific energy and obtaining superior surface roughness. However, semi-synthetic oil was favourable for obtaining better tool life.

Although vegetable-based cutting oils have many advantages, some researchers have found that synthetic oils work better as cutting fluids. Kazeem et al. (2020) found that at some cutting parameters, mineral-based oil provided better surface finish than jatropha oil-based cutting fluid. Considering this, several initiatives have been undertaken to enhance the effectiveness of vegetable-based oils. One approach involved incorporating appropriate nanoparticles into the vegetable oil. These nanoparticles help lower friction at the contact interface by enhancing the cutting fluid's thermal conductivity and viscosity. Singh et al. (2021) attempted to improve the lubrication ability of jatropha oil by dispersing varying concentrations of Al_2O_3 and ZrO nanoparticles. They observed that the most preferred viscosity was obtained by dispersing 1 wt.% of Al_2O_3. However, the best thermal properties were obtained by dispersing 0.5 wt.% of ZrO. It is evident from the literature that nanoparticle mixed vegetable oils have a great potential as a cutting fluid. Several vegetable oils such as jatropha, neem, coconut, sunflower and castor oils have been studied in machining research. However, the oil extracted from the seeds of Putranjiva roxburghii, a medicinal plant of Southeast Asia, has not been studied as a cutting fluid so far. This non-edible vegetable oil belongs to the Putranjiva roxburghii plant of the family Euphorbiaceae. This plant is also referred to as Drypetes roxburghii. It is abundant in certain reasons and can be an economical alternative to existing cutting fluids.

This chapter explores the development of neat Putranjiva roxburghii oil-based nanofluids through the addition of nanoparticles and evaluates their thermophysical, lubricating, and other relevant properties when used as cutting fluids in machining applications. The methodology for preparing the nanofluids and the characterization techniques are discussed in detail in Section 3.3.5 in Chapter 3. The remaining sections of this chapter are organized as follows: Section 4.2 discusses the concentrations of nanoparticles used for the preparation of nanofluids. Section 4.3 presents the characterization techniques applied to both the nanoparticles and the prepared nanofluids. Section 4.4 provides the results and discussion of the experimental findings. Section 4.5 covers the optical study of the nanofluids. Section 4.6 offers a comparison between the prepared nanofluids and other commonly used cutting fluids. Finally, Section 4.7 presents a summary of the key findings from this chapter.

4.2 Physical Propertiies of Neat Putranjiva Roxburghi oil

The density of pure Putranjiva roxburghii oil were measured at two different temperatures, viz., 20 °C and 40 °C. The obtained results for density were 912.2 kg/m^3 and 900.1 kg/m^3, respectively. In other investigations, Singh et al. (2022) and Subramanian et al. (2015) reported the density of Putranjiva roxburghii oil to be 913 kg/m^3 and 923 kg/m^3, respectively. Both values were taken at a temperature of 15 °C. The pour point of the fluid represents the lowest temperature at which it loses its fluidity. The cloud point refers to the temperature at which a fluid, begins to show cloudy or hazy appearances. The pour point of the Putranjiva roxburghii oil was obtained as –7 °C and the cloud point was obtained as –2 °C. Considering it in extremely low weather conditions, the free flow of a cooling fluid under low temperatures is important. In machining processes, a lubricant must effectively dissipate heat from the cutting area. Therefore, it should possess high flash and fire points to minimize the risk of ignition. When a fluid is exposed to an external flame or heat source, it can evaporate and ignite at the lowest temperature known as the flash point. The fire point is the temperature at which the vapors produced by the fluid are able to sustain a continuous flame following ignition. In this research, the flash point and fire point of Putranjiva roxburghii oil were determined to be 260 °C and 280 °C, respectively.

4.3 Morphological, Structural and Optical Analysis of Nanoparticles and Putranjiva Roxburghii oil based Nanofluids

4.3.1 Morphological study of nanoparticles

To examine the structure and surface features of the nanoparticles, Field Emission Scanning Electron Microscopy (FESEM) was used to observe their shape, size, and distribution.

Field Emission Scanning Electron Microscopy (FESEM)

Figures 4.1 (a) and 4.1 (c) depict the shape of SiO_2 and graphene nanoparticles, respectively. It is observed that SiO_2 have spherical shapes while graphene nanoparticles have a flower shape. Figures 4.1 (b) and 4.1 (d) reflect the EDX spectra of SiO_2 and graphene nanoparticles, respectively. An EDX analysis of SiO_2 nanoparticles shows 55.3 wt.% of oxygen, 30.7 % of silicon, 14 % of carbon and the rest are impurities. An EDX analysis of graphene nanoparticles shows 99.9% purity of carbon. Nanoparticles are said to be devoid of any impurity elements based on their elemental makeup.

FIGURE 4.1 FESEM and EDX analysis: (a, b) SiO_2 nanoparticles and (c, d) graphene nanoparticles

4.3.2 Structural study of nanoparticles

The structural analysis of nanoparticles, including their crystalline nature and phase composition, was carried out using X-ray Diffraction (XRD) techniques.

X-ray Diffraction (XRD) analysis

The crystal structure, lattice parameters and phases of SiO_2 and graphene nanoparticles were determined. The intensity of the radiation source approached 4,000 counts per second. The analysis was done in the range $10° < 2\theta < 80°$. Figures 4.7 (a) and (b) show the XRD patterns of SiO_2 and graphene nanoparticles, respectively. The spectrum in Figure 4.2 (a) confirmed the amorphous pattern of SiO_2 nanoparticles. The wide peak has a structure that is unstructured. There is a sharp peak (011) at 28.22°. It is the peak for crystalline SiO_2. The scores were verified against the inorganic crystal structure database (ICSD) reference code 96-901-2605 file, the SiO_2 has no impurities peak.

The graphene nanoparticles had a hexagonal crystalline structure. There is a sharp peak (002) at about 26.23° as shown in Figure 4.2 (b). Based on Bragg's law, the (002) plane has a *d*-spacing of 3.395 Å. Other peak (011) is seen at 44.36°. Since graphene is a 2-D material with a hexagonal lattice structure, it does not show a complicated XRD pattern like 3-D crystals do. The result was verified with ICSD reference code 96-101-1061. Similar observation regarding XRD pattern of graphene is also reported by Rahimi et al. (2015).

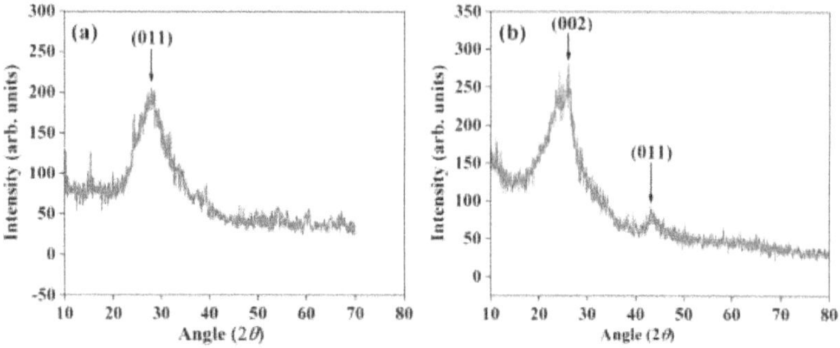

FIGURE 4.2 XRD patterns: (a) SiO_2 nanoparticles and (b) graphene nanoparticles

4.3.3 Optical analysis of nanoparticles and nanofluids

An optical study refers to the analysis of a material's behaviour and properties when it interacts with light. Fourier Transform Infrared Spectroscopy (FTIR) identified the functional groups, confirming the chemical integrity and successful formation of the nanoparticles. In the context of nanofluids or nanoparticles, an optical study usually involves techniques that observe how these fluids absorb, transmit, reflect, or scatter light.

Fourier Transform Infrared Spectroscopy (FTIR) of nanoparticles

The Fourier Transform Infrared Spectroscopy (FTIR) results for SiO_2 and graphene nanoparticles are displayed in Figures 4.3 (a) and (b), respectively. In the case of SiO_2, the absorption band at 1073.34 cm^{-1} corresponds to the asymmetric stretching vibration of Si–O–Si. The peak at 795.83 cm^{-1} and 615.41 cm^{-1} represents the bending vibration of Si–O. Similar FTIR results for SiO_2 have also been reported by Mohammadi et al. (2019). For graphene nanoparticles as shown in Figure 4.3 (b), the peak at around 3472.52 cm^{-1} may indicate the presence of O-H stretching vibrations. It can be associated with the adsorption of water or oxygen-containing functional groups on the graphene surface. Other peaks in the range of 2124.59 cm^{-1} are shown for C-H stretching vibrations. They vary depending on the specific chemical functionalization of the graphene. Graphene is primarily composed of carbon atoms bonded in a hexagonal lattice. The sharp peak with a wave number of 1637.48 cm^{-1} shows the presence of graphene, corresponding to C=C stretching vibrations in the carbon-carbon double bonds. Various oxygen configurations in the structure include the vibration modes shown on peak 1416.58 cm^{-1} and 1112.54 cm^{-1} due to the presence of epoxy or hydroxyl functional groups in graphene nanoparticles. Similar observations were also reported by Kumar et al. (2011).

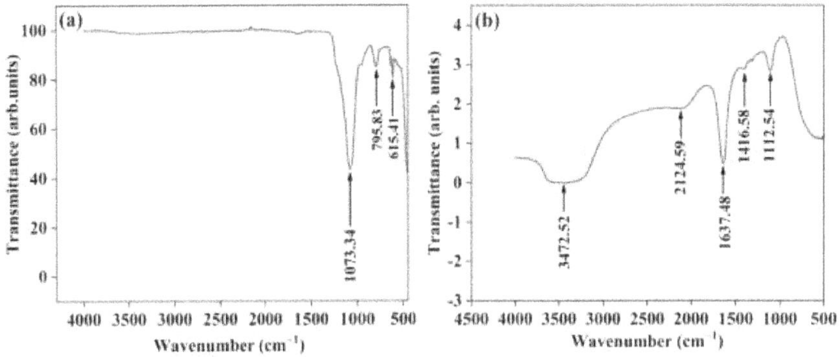

FIGURE 4.3 FTIR peaks: (a) SiO_2 nanoparticles and (b) graphene nanoparticles

Fourier Transform Infrared Spectroscopy (FTIR) of nanofluids

Figures 4.4 (a), (b), and (c) display the FTIR spectrum of the different wt. % (0.25%, 0.5%, 0.75%, and 1%) of SiO_2 and graphene nanoparticles dispersed Putranjiva roxburghii oil based nanofluids and hybrid nanofluids. The absorbance range was considered between 500 to 4000 cm^{-1}. The prominent peaks observed in the spectrum correspond to wavenumbers 3330, 2922, 2853, 2424 and 1742 cm^{-1}. They can be attributed to the asymmetric stretching of the Putranjiva roxburghii oil link as well as the Si-O-Si bond. Additionally, the peaks observed at wavenumbers 1460, 1212, 1160 and 720 cm^{-1} are indicative of Si–O vibrations inside the bio-oil. It is likely caused by the presence of nanoparticles. These frequencies match the bonding structure of O-Si-O as reported by Nguele et al. (2019). Due to this bonding structure, the transmittance is shifted accordingly which is lower in SiO_2 and hybrid nanofluids while higher in the case of graphene dispersed nanofluids. This particular arrangement can vibrate in three different ways, viz., by bending, by stretching in a symmetric way and by stretching in an uneven way. This shows that the samples prepared had a clean structure.

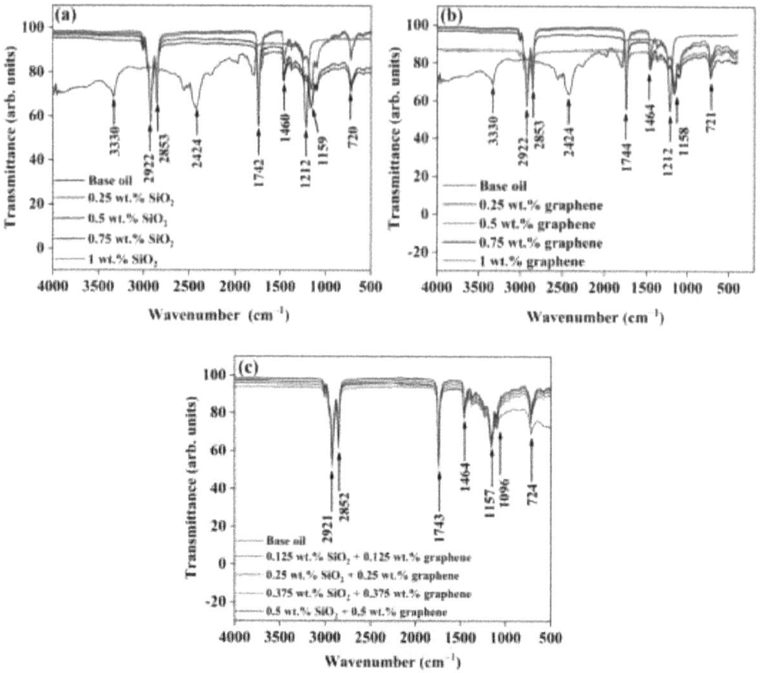

FIGURE 4.4 FTIR peaks: (a) SiO_2 based nanofluid, (b) graphene based nanofluid and (c) SiO_2 and graphene based hybrid nanofluid

UV-Vis study of nanofluids

The Ultraviolet-Visible (UV-Vis) spectrum for graphene and SiO_2-based nanofluids with different wt.% are shown in Figure 4.5. All of the chemicals that have been looked at have a maximum absorbance peak between 410 and 680 nm. The highest and minimum values of the absorption peaks are around 672 nm. In the hybrid nanofluid shown in Figure 4.10 (c), the different peaks are reflected at 339, 414, 481, and 674 nm. Absorbance varies only due to C=C bonds. In another investigation Mohammadi et al. (2019), the mass concentration of carbon fiber-reinforced graphene nanoplatelets has been reported. In this study, as the concentrations of nanoparticles increase, the absorbance steadily rises, displaying minimal changes in the case of SiO_2 and hybrid nanofluids, but a significant increase in the case of graphene nanofluids.

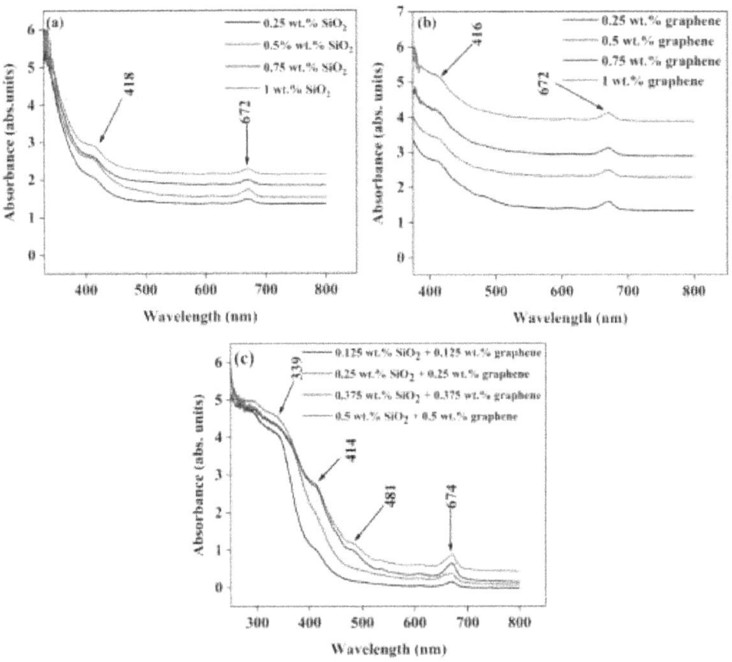

FIGURE 4.5 UV-Vis analysis: (a) SiO_2 based nanofluid, (b) graphene based nanofluid and (c) SiO_2 and graphene based hybrid nanofluid

4.4 Thermal, Rheological and Wettability Analysis of Prepared Nanopfluids

This section outlines and analyzes the comprehensive findings from the physico-chemical and structural evaluation of the synthesized nanofluids. Essential thermophysical attributes such as thermal conductivity, specific heat capacity, density and viscosity were determined to assess the heat transfer capabilities and flow characteristics of the formulated nanofluids. pour point, cloud point, flash point, and fire point analyses were also conducted to determine the low- and high-temperature operating limits and safety characteristics of the fluids. These properties are crucial for assessing the suitability of the nanofluids in machining applications involving elevated temperatures.

The FTIR analysis of the nanofluids further confirmed the compatibility between the base emulsion and the dispersed nanoparticles. Rheological studies were carried out to investigate the flow behavior under varying shear conditions, offering insights into their Newtonian or non-Newtonian nature. The UV-Visible spectroscopy (UV-Vis) analysis was conducted to assess the optical absorbance and dispersion stability of the nanofluids over time. Furthermore, wettability tests (contact angle measurements) were performed to evaluate the spreading behavior of the nanofluids on metallic surfaces, which directly influences lubrication efficiency during machining. Together, these findings offer a thorough comprehension of the thermal, chemical, and physical behavior of the developed nanofluids, enabling their suitability to be judged for sustainable and high-performance metal cutting operations. The developed composite is a favorable candidate for utilization in solar water heating systems.

4.4.1 Thermal conductivity and volumetric heat capacity

The thermal conductivity and volumetric heat capacity of nanofluids with various weight concentrations (0%, 0.25%, 0.5%, 0.75% and 1%) containing SiO_2 and graphene nanoparticles were obtained as shown in Figures 4.1 and 4.2, respectively. Here, 0% refers to pure Putranjiva roxburghii oil. The thermal conductivity and volumetric heat capacity of pure oil were obtained as 0.158 W/m.K at and 2.471 MJ/m^3.K at 30 °C, respectively.

The size difference between SiO_2 particles (30–50 nm) and graphene sheets (3–6 nm) strongly affects the thermal conductivity of the nanofluids. Smaller graphene particles offer higher surface area and shorter heat-transfer paths, improving thermal transport through enhanced Brownian motion and interfacial interactions. Their naturally high thermal conductivity also contributes to greater heat-transfer capability. Larger SiO_2 particles, with lower surface area and limited phonon mobility, provide comparatively weaker thermal conductivity enhancement (Angayarkanni et al., 2015).

Figures 4.6 and 4.7 show that the presence of the nanoparticles enhances thermal conductivity and volumetric heat capacity. However, the enhancement in thermal conductivity is less than that in volumetric heat capacity. For SiO_2 nanofluid, the highest value of thermal conductivity is 0.168 W/m.K at SiO_2 concentration of 0.5% and 1%; however, it is only slightly lower (0.167 W/m.K) at SiO_2 concentration of 0.75%. The highest value of volumetric heat capacity was 2.809 MJ/m^3.K at SiO_2 concentration of 0.75%.

In case of graphene nanoparticles in the base oil, the value of thermal conductivity reached a value of 0.174 W/m.K at a concentration of 0.75%. This is the highest value amongst all combinations. At 0.75% of graphene nanoparticles, the volumetric heat capacity is obtained as 3.299 MJ/m^3.K. Figure 4.6 (c) presents the thermal conductivity of hybrid nanofluids, highlighting a peak value of 0.171 W/m·K at a 1% concentration. The maximum volumetric heat capacity among all mixtures, recorded as 3.784 MJ/m^3·K, occurred at a 0.75% concentration, also illustrated in Figure 4.7 (c).

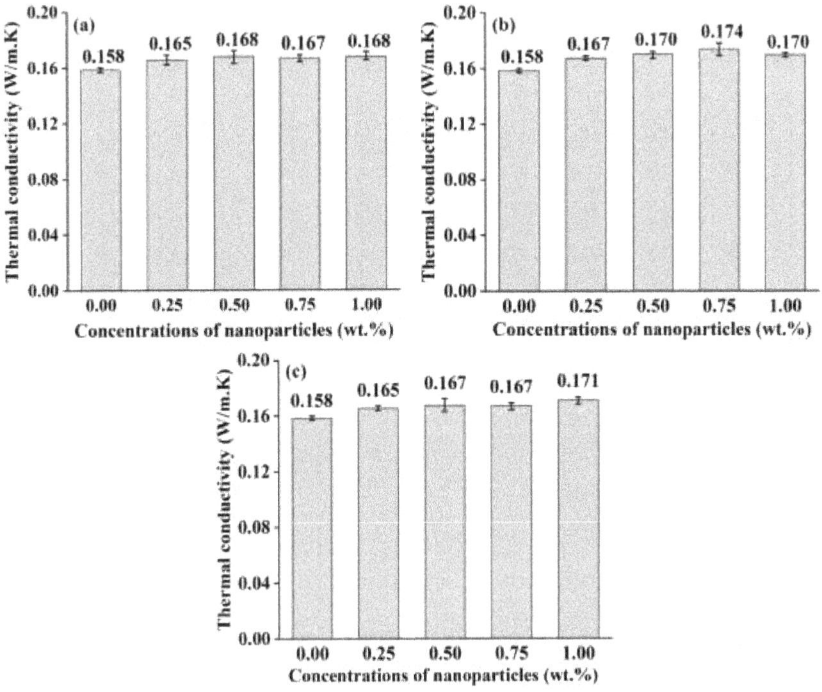

FIGURE 4.6 Variations of thermal conductivity with different concentrations of nanoparticles: (a) SiO_2, (b) graphene and (c) SiO_2 + graphene.

Nanofluids containing graphene nanoparticles demonstrate the greatest thermal conductivity, which is attributed to the exceptional thermal conductivity properties of graphene (Kole and Dey, 2013). The increased thermal conductivity seen in nanofluids is partly due to Brownian motion and the development of liquid layers surrounding the particles at the liquid-particle boundary. In the context of Brownian motion, nanoparticles traverse through the liquid medium and have the potential to collide with one another, facilitating direct heat transfer between solid particles.

For graphene-based and SiO_2-based nanofluids, the slight reduction in thermal conductivity and volumetric heat capacity above 0.75 wt.% is mainly due to agglomeration and loss of dispersion stability at higher particle loadings. Excess nanoparticles cluster together, increasing the effective particle size, limiting Brownian motion, and restricting heat transfer pathways, which also raises viscosity and reduces overall heat-carrying ability.

In contrast, the hybrid nanofluid shows improved thermal conductivity and volumetric heat capacity beyond 0.75 wt.% because the combination of graphene and SiO_2 provides better dispersion stability and limits agglomeration. Their complementary interaction promotes a more uniform suspension and efficient energy transport, allowing the hybrid fluid to maintain enhanced thermal performance at higher concentrations.

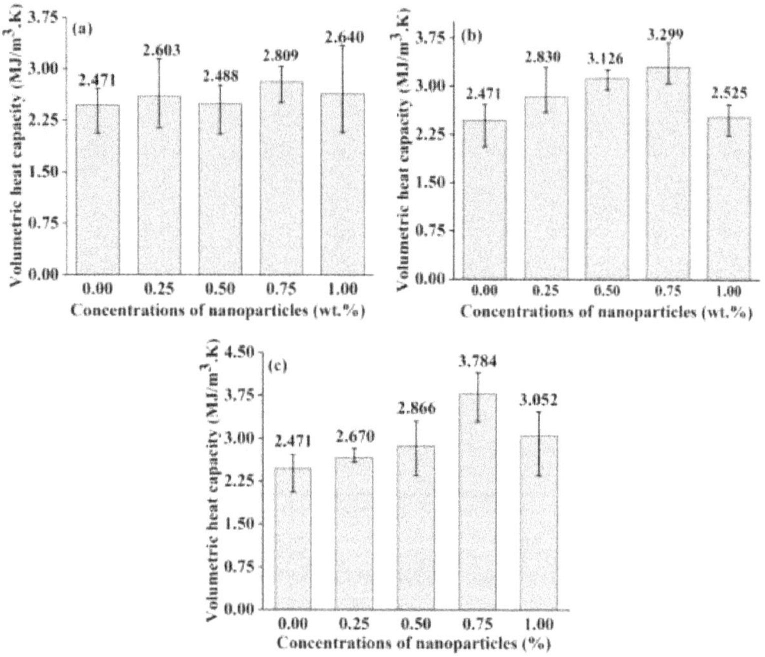

FIGURE 4.7 Variations of volumetric heat capacity with different concentrations of nanoparticles: (a) SiO_2, (b) graphene and (c) SiO_2 + graphene. Error bars are based on five replicates.

This phenomenon ultimately leads to an increase in the thermal conductivity of nanofluids (Ambreen and Kim 2020). As per observations indicated by Arshad et al. (2019), the ability of nanofluids to transfer heat depends on their concentration, particle size and shape, material makeup, level of purity, motion and temperature. Increasing the concentration beyond a limit, hampers the free motion of particles and causes agglomeration, thus reducing the thermal conductivity. Enhancement of volumetric heat capacity with increase in the concentration of nanoparticles is due to enhanced heat capacity of nanoparticles, but it is not properly understood why the maximum value is attained at 0.75% concentration in the hybrid nanofluid.

4.4.2 Rheological study of nanofluids

Rheological experiments have been conducted to examine the response of the fluid to shear deformation. Figures 4.8–4.10 illustrate how the viscosity of Putranjiva roxburghii oil behaves under different conditions in the temperature range 30 °C to 120 °C. At 60 °C, the viscosity of Putranjiva roxburghii oil reduces with an increase in the shear rate, indicating a shear-thinning property within the fluid. Viscosity strongly influences the lubrication, flow behaviour, and cooling performance of nanofluids. Higher viscosity helps form a stable lubricating film and increases load-carrying ability, while very high viscosity can obstruct fluid entry into the cutting zone and reduce heat removal. Lower viscosity improves penetration and cooling but may weaken lubrication. Thus, viscosity affects machining stability, chip formation, and surface quality (Hirudayanathan et al. 2023).

Rheological analysis is essential because viscosity determines how efficiently the nanofluid can reach the tool–workpiece interface, form a protective layer, dissipate heat, and maintain nanoparticle dispersion under machining conditions. This characterization ensures that the chosen nanofluid behaves suitably in high-temperature, high-shear environments.

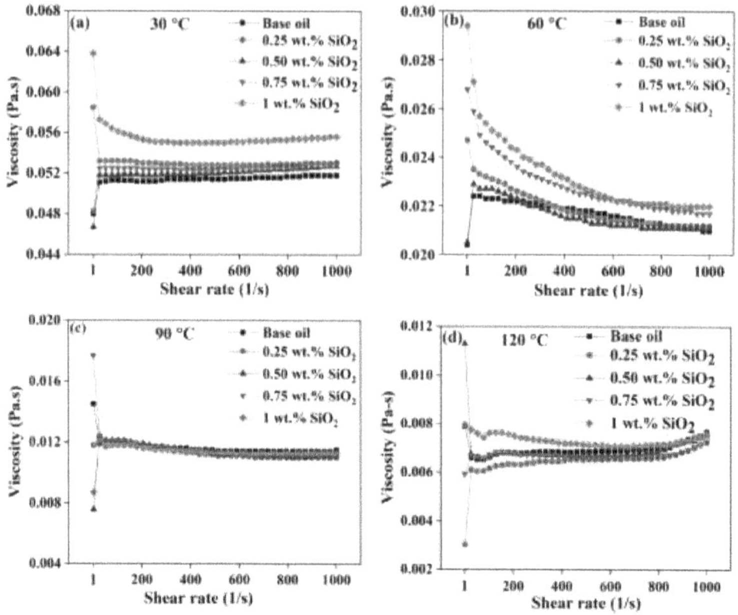

FIGURE 4.8 Variations of viscosity with shear rate in SiO_2 nanoparticles dispersed nanofluids at different temperatures: (a) 30 °C, (b) 60 °C, (c) 90 °C and (d) 120 °C

However, there is no significant dependence of viscosity on shear rate for other temperatures. The introduction of nanoparticles causes an increase in the viscosity of Putranjiva roxburghii oil. This increase in viscosity is due to the temporary structural adjustments in the fluid molecules during rotation of the concentric assembly of the rheometer apparatus. Graphene-based nanofluids exhibit a more substantial enhancement in viscosity when compared to SiO_2-based and hybrid nanofluids as shown in Figure 4.9. This can be attributed to differences in particle size and nanoparticle characteristics. Nanofluids exhibit similar shear-thinning behavior as observed in pure Putranjiva roxburghii oil.

The variation of viscosity with shear rate in graphene and hybrid nanoparticles dispersed nanofluids are depicted in Figures 4.9 and 4.10. When the spindle in the rheometer setup turns, the fluid molecules rearrange themselves for a short time. The molecules of a fluid line up in the direction of the shear. This reduces the viscosity. When nanoparticles are added to Putranjiva roxburghii oil, it become thicker.

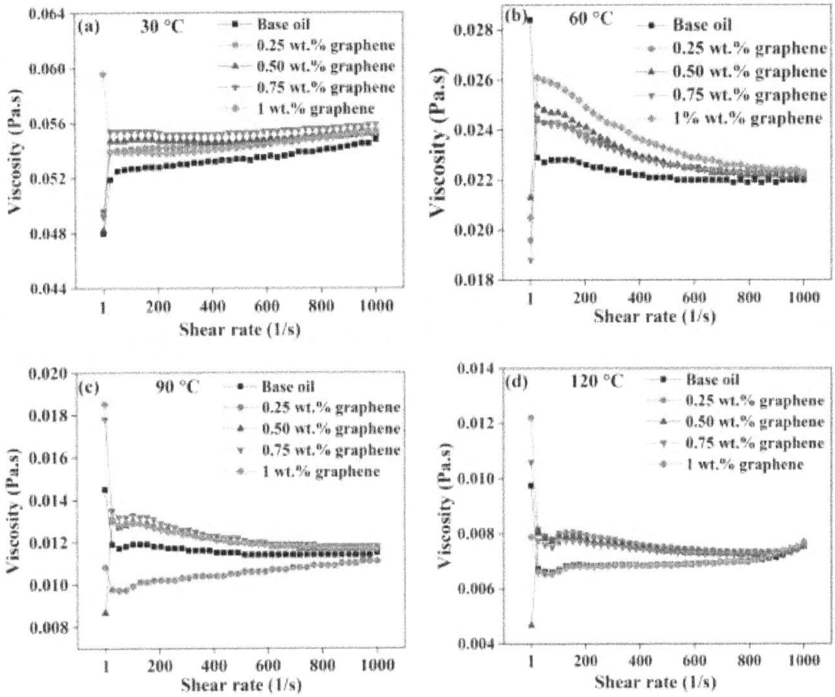

FIGURE 4.9 Variations of viscosity with shear rate in graphene nanoparticles dispersed nanofluids at different temperatures: (a) 30 °C, (b) 60 °C, (c) 90 °C and (d) 120 °C

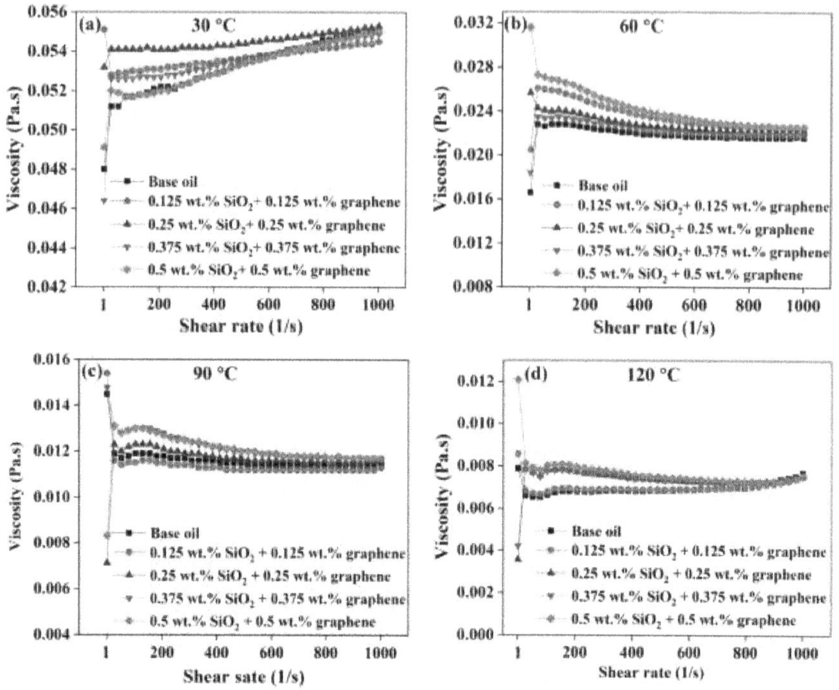

FIGURE 4.10 Variations of viscosity with shear rate in SiO_2 and graphene nanoparticles dispersed hybrid nanofluids at different temperatures: (a) 30 °C, (b) 60 °C, (c) 90 °C and (d) 120 °C

4.4.3 Wettability study of nanofluids

To understand how well a liquid spreads on a solid surface, the angle is measured between the liquid and solid surfaces at the location where they meet. When the droplet spreads more on the surface, it means it can provide better lubrication. The extent of the droplet spreading depends on how the liquid tries to reduce the energy of the interface between the liquid and the solid. This energy is related to surface tension that is influenced by van der Walls forces, especially at very small distances, between 1–100 nanometres (Singh et al.2021; Khandekar et al. 2012).

Incorporating nanoparticles into the fluid changes the free energy dynamics at the boundary between the liquid and solid surfaces. This change occurs because nanoparticles affect the liquid that sticks to the solid surface. The results are displayed in Figures 4.11–4.14. It is observed that adding nanoparticles reduces the contact angle. For instance, Putranjiva roxburghii oil has an average contact angle of around 34° as shown in Figure 4.11, whereas various nanofluids concentrations result in smaller contact angles. This reduction occurs because the molecules of the droplets gain more kinetic energy and experience brownian motion due to the presence of nanoparticles in the fluid. This results in the molecules moving away from the fluid-solid interface, which lowers the contact angle.

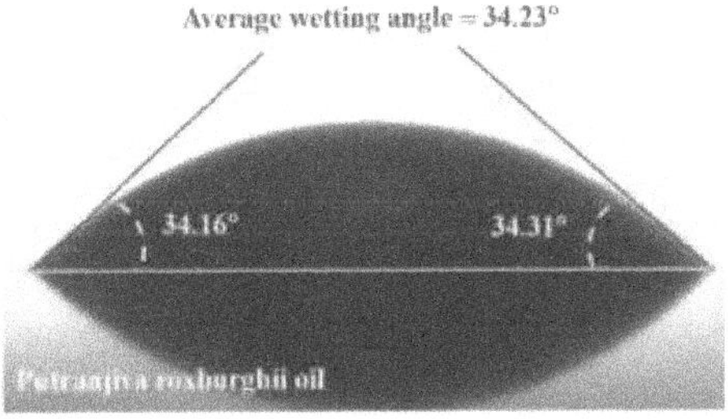

FIGURE 4.11 Contact angle of Putranjiva roxburghii oil

Figure 4.11 presents the results of contact angles for various SiO_2-based nanofluids concentrations. The lowest contact angle, measuring 26.64°, was observed at a concentration of 0.5%. At a concentration of 0.75%, the wetting angle increases, but it subsequently decreases at the concentration of 1%. In the context of graphene-based nanofluids shown in Figure 4.13, as the nanoparticle concentration rises, wetting angles show a declining tendency. It reaches an optimal value of 24.19° degrees at a concentration of 0.75%. However, when concentrations rise to 1%, there is an increase in the contact angle. This can be attributed to the increased viscosity of the nanofluids.

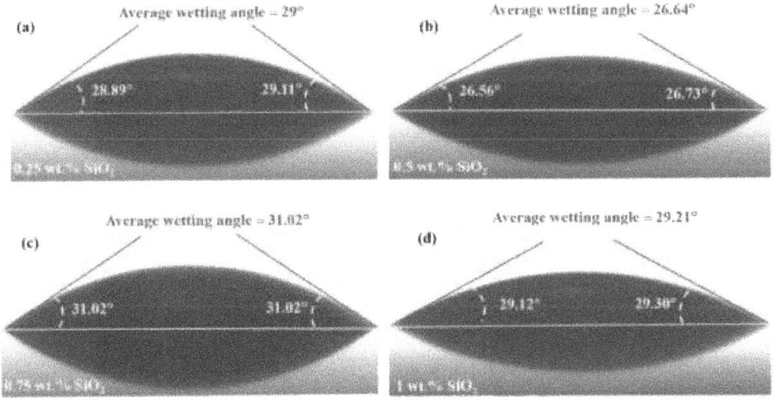

FIGURE 4.12 Contact angle of different wt.% of SiO_2 nanoparticles dispersed nanofluids: (a) 0.25%, (b) 0.5%, (c) 0.75% and (d) 1%

Furthermore, graphene-based nanofluids exhibit a more noticeable drop in contact angle than SiO_2-based nanofluids. However, in Figure 4.14, hybrid nanofluids show a slightly smaller reduction in contact angle.

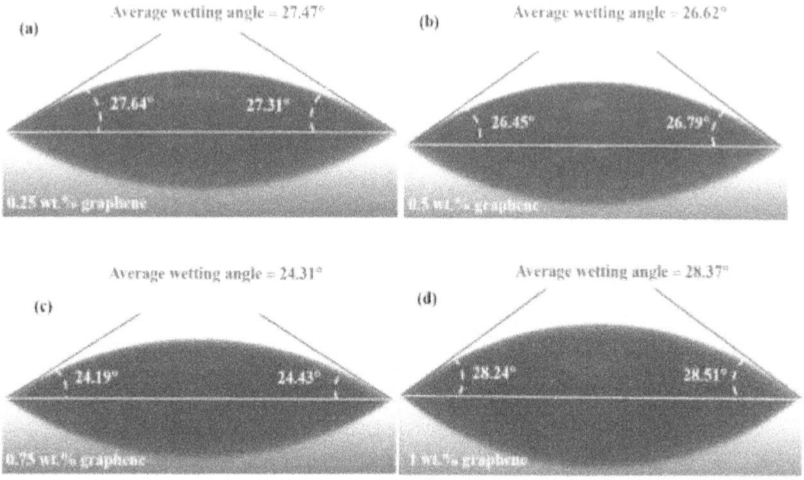

FIGURE 4.13 Contact angle of different wt.% of graphene nanoparticles dispersed nanofluids: (a) 0.25%, (b) 0.5%, (c) 0.75% and (d) 1%

Increasing the concentration of nanoparticles reduces the contact angle up to a certain limit. Due to variations in their physical and chemical properties, different nanoparticles have distinct limits.

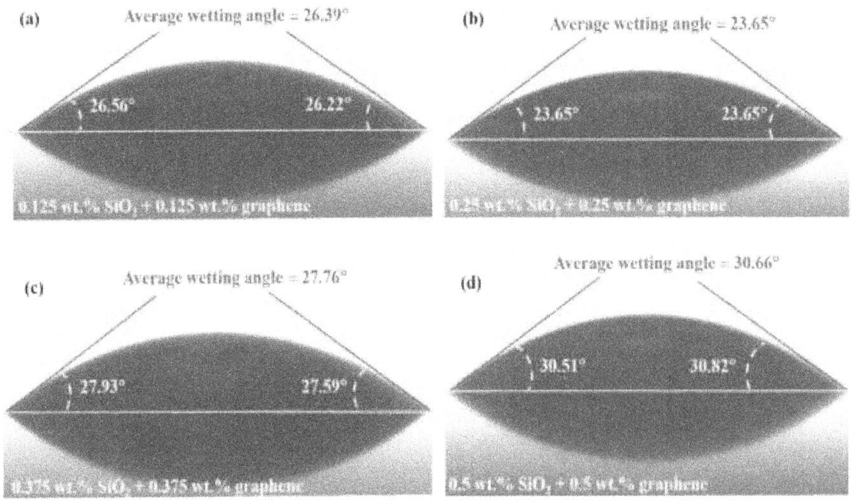

FIGURE 4.14 Contact angle of different wt.% of graphene plus SiO_2 nanoparticles dispersed nanofluids: (a) 0.25%, (b) 0.5%, (c) 0.75% and (d) 1 %

Beyond a certain limit, increasing the nanoparticle concentration leads to the clumping together of nanoparticles in the nanofluids, resulting in an increase in the contact angle. For example, at a 0.75 wt.% concentration of unitary nanoparticles, a 28.98% reduction in contact angle is observed for graphene-based nanofluids as compared to base oil. However, for SiO_2 based nanofluid, a 22.17% reduction in the contact angle was observed at a concentration of 0.5%, as shown in Figure 14. In addition, a 30.9% reduction is observed in the case of hybrid nanofluids with a 0.5 wt.% concentration of graphene plus SiO_2-based nanoparticles when compared to Putranjiva roxburghii oil. In another study by Gajrani et al. (2019), a similar reduction in the contact angle was obtained in terms of unitary and hybrid nanofluids.

4.5 Comparison of Prepared Nanofluids with Different Cutting Fluids

The properties of Putranjiva roxburghii oil are compared with water and mineral oil. Water is considered as an ideal coolant, but is not used directly as the cutting fluid due to corrosion problem. Mineral oils are used straight or with additives but are not environmentally benign. Properties of water and mineral oil are collected from the literature are shown in Table 4.2. The thermal conductivity of Putranjiva roxburghii oil is approximately one-quarter that of water, yet it surpasses mineral oil by 1.5 times. While its volumetric heat capacity is notably greater than that of mineral oil, it remains lower compared to water.

Viscosity of Putranjiva roxburghii oil is comparable with that of the mineral oil. Its density is in between the densities of water and mineral oil. Although pour and cloud points of Putranjiva roxburghii oil are greater than those of mineral oils, there are nonetheless in the sub-zero range. Its flash and fire points are slightly greater than those of mineral oil. The low contact angle of Putranjiva roxburghii oil on Inconel 718 surface indicates high wettability, which is a positive feature of a cutting fluid. Overall, the lubricating properties (viscosity and contact angle) of Putranjiva roxburghii oil appear to be better than those of water and mineral oils. As a coolant, Putranjiva roxburghii oil has better properties than those of mineral oils, although they are no match to water. Nanoparticle addition can improve Putranjiva roxburghii oil's qualities even more.

TABLE 4.1 Thermophysical and wettability properties of different cutting fluids

Properties	**Water**	**Mineral oil**	**Putranjiva oil**
Thermal conductivity (W/m.K)	0.615 at 30 °C [16]	0.1056 at 25 °C [21]	0.158 at 30 °C
Volumetric heat capacity (MJ/m^{3}K)	4.186 [17]	1.39 [22] (specific heat is multiplied with density)	2.471 at 30 °C
Viscosity (Pa.s)	0.0008 at 30 °C 0.0005 at 60 °C [18]	0.025–0.08 at 20 °C [23]	0.048 at 30 °C 0.022 at 60 °C
Density (kg/m^3)	998.29 at 20 °C 992.25 at 40 °C [19]	833 at 25 °C [23]	912.2 at 20 °C 900.1 at 40 °C
Pour point (°C)	–	–30 °C [24]	–7 °C
Cloud point (°C)	–	–6 °C [24]	–2 °C
Flash point (°C)	–	230 °C [25]	260 °C
Fire point (°C)	–	260 °C [25]	280 °C
Contact angle (°)	50°–80° on metallic surfaces [20]	55.2° on AISI steel [15]	34° on Inconel 718 surface

4.6 Summary

This chapter explored the lubrication ability of Putranjiva roxburghii oil. By considering this as a base oil, nanofluids are developed by adding different concentrations of graphene and SiO_2 nanoparticles. The microstructural analysis validated the morphology and dimensions of the nanoparticles as supplied. A brief examination indicates that a 0.75% nanoparticle concentration in the base oil yields the best balance of properties. This non-edible oil is one of the least explored vegetable oils. Experimental findings of the developed nanofluids show some promising scopes in the area of lubrication. The findings of the study are as follows:

- Addition of SiO_2 and graphene nanoparticles enhanced the thermal conductivity of the nanofluid. The highest value of thermal conductivity was obtained for 0.75 wt.% of graphene-dispersed nanofluids.
- Nanofluids cause some enhancement in viscosity, which provides better lubricity.
- Wettability studies conducted in the Inconel 718 material revealed that the addition of nanoparticles lowered the contact angle. Graphene plus SiO_2 hybrid nanofluids showed the most desirable wettability characteristics. As a base oil also, Putranjiva roxburghii oil has better lubricity than that of a mineral oil.

Overall, the developed Putranjiva roxburghii based nanofluids can be used as a cutting fluids in different machining operations. It is benign to environment as well as the health of the operator. As of now, the oil is not very expensive; Putranjiva plants are found abundantly in certain regions and grow naturally. Hence, use of Putranjiva roxburghii based nanofluids is a sustainable move.

References

Ambreen, T. and Kim, M.H., 2020. Influence of particle size on the effective thermal conductivity of nanofluids: A critical review. Applied Energy, 264, p.114684.

Angayarkanni, S.A., Sunny, V. and Philip, J., 2015. Effect of nanoparticle size, morphology and concentration on specific heat capacity and thermal conductivity of nanofluids. *Journal of nanofluids*, *4*(3), pp.302-309.

Arshad, A., Jabbal, M., Yan, Y. and Reay, D., 2019. A review on graphene based nanofluids: Preparation, characterization and applications. Journal of Molecular Liquids, 279, pp.444–484.

Cloud Point, 2007, <https://www.sciencedirect.com/topics/chemistry/cloud-point Accessed on December 24>, 2023

Dixit, U.S., Sarma, D.K. and Davim, J.P., 2012. Environmentally friendly machining. Springer, New York.

Fernandes, P. and Prabhu, K.N., 2008. Comparative study of heat transfers and wetting behaviour of conventional and bioquenchants for industrial heat treatment. International Journal of Heat and Mass Transfer, 51(3-4), pp.526–538.

Gajrani, K.K., Suvin, P.S., Kailas, S.V. and Mamilla, R.S., 2019. Thermal, rheological, wettability and hard machining performance of MoS_2 and CaF_2 based minimum quantity hybrid nano-green cutting fluids. Journal of Materials Processing Technology, 266, pp.125–139.

Gupta, M.K. and Korkmaz, M.E., 2023. A conceptual framework for sustainability impact assessment in machining bohler tool steel under hBN-enriched nano cutting fluids environment. Sustainable Materials and Technologies, 37, p.e 00669.

Hirudayanathan HP, Debnath S, Anwar M, Johar MB, Elumalai NK, Mohammed Iqbal U. A review on influence of nanoparticle parameters on viscosity of nanofluids and machining performance in minimum quantity lubrication. *Proceedings of the Institution of Mechanical Engineers, Part E: Journal of Process Mechanical Engineering*. 2023;239(2):1005-1024.

Hu, X., Yin, D., Chen, X. and Xiang, G., 2020. Experimental investigation and mechanism analysis: Effect of nanoparticle size on viscosity of nanofluids. Journal of Molecular Liquids, 314, p.113604.

Kazeem, R.A., Fadare, D.A., Abutu, J., Lawal, S.A. and Adesina, O.S., 2020. Performance evaluation of jatropha oil-based cutting fluid in turning AISI 1525 steel alloy. CIRP Journal of Manufacturing Science and Technology, 31, pp.418–430.

Khandekar, S., Sankar, M.R., Agnihotri, V. and Ramkumar, J., 2012. Nano-cutting fluid for enhancement of metal cutting performance. Materials and manufacturing processes, 27(9), pp.963–967.

Kole, M. and Dey, T.K., 2013. Investigation of thermal conductivity, viscosity, and electrical conductivity of graphene based nanofluids. Journal of Applied Physics, 113(8), 084307.

Kumar, N., Das, S., Bernhard, C. and Varma, G.D., 2013. Effect of graphene oxide doping on superconducting properties of bulk MgB2. Superconductor Science and Technology, 26(9), p.095008.

Kuram, E., Ozcelik, B., Bayramoglu, M., Demirbas, E. and Simsek, B.T., 2013. Optimization of cutting fluids and cutting parameters during end milling by using D-optimal design of experiments. Journal of Cleaner Production, 42, pp.159–166.

Merck, <https://www.sigmaaldrich.com/IN/en/product/sial/330779#product-documentation> Accessed on December 22, 2023

Mohammadi, M., Khorrami, M.K. and Ghasemzadeh, H., 2019. ATR-FTIR spectroscopy and chemometric techniques for determination of polymer solution viscosity in the presence of SiO2 nanoparticle and salinity. Spectrochimica Acta Part A: Molecular and Biomolecular Spectroscopy, 220, p.117049.

Nguele, R., Sreu, T., Inoue, H., Sugai, Y. and Sasaki, K., 2019. Enhancing oil production using silica-based nanofluids: preparation, stability, and displacement mechanisms. Industrial & Engineering Chemistry Research, 58(32), pp.15045–15060.

Rahimi, R., Zargari, S., Yousefi, A., Berijani, M.Y., Ghaffarinejad, A. and Morsali, A., 2015. Visible light photocatalytic disinfection of E. coli with TiO2–graphene nanocomposite sensitized with tetrakis (4-carboxyphenyl) porphyrin. Applied Surface Science, 355, pp.1098–1106.

Singh, Y., Singh, N.K., Sharma, A. and Badhotiya, G.K., 2022. Tribo-mechanism analysis of epoxidized Putranjiva oil (Putranjiva roxburghii) and understanding lubricity with silicon oxide nanoparticle addition. Proceedings of the Institution of Mechanical Engineers, Part J: Journal of Engineering Tribology, 236(4), pp.648–658.

Somlyai-Sipos, L. and Baumli, P., 2022. Wettability of metals by water. Metals, 12(8), p.1274.

Subramanian, N., Mahendradas, D.K., Kasirajan, R. and Sahadevan, R., 2015. Bio-oil separation from potential non-edible urban waste source Putranjiva roxburghii. Separation Science and Technology, 50(13), pp.2066-2074.

The Engineering Toolbox, 2004, <https://www.engineeringtoolbox.com/water-dynamic-kinematic-viscosity-d_596.html> Accessed on December 24, 2023

The Engineering Toolbox, 2004, <https://www.engineeringtoolbox.com/specific-heat-fluids-d_151.html> Accessed on December 22, 2023

The Engineering Toolbox, 2018, <https://www.engineeringtoolbox.com/water-liquid-gas-thermal-conductivity-temperature-pressure-d_2012.html> Accessed on 24 December, 2023.

Valves Instrument Plus Ltd. <file:///C:/Users/hp/Downloads/Density_Of_Water_Tables.pdf > Accessed on December 23, 2023

Volumetric Heat Capacity, 2017, <https://www.sciencedirect.com/topics/engineering/volumetric-heat-capacity> Accessed on 24 December, 2023.

Xiang, D., Shen, L. and Wang, H., 2019. Investigation on the thermal conductivity of mineral oil-based alumina/aluminum nitride nanofluids. Materials, 12(24), p.4217.

CHAPTER

5

Milling Performance of Inconel 718 under Dry, ir-Cooling, and Bio-oil Based cooling with MQL Technique

5.1 Introduction

In recent times, industries have started using hard materials such as AISI H13 steel, titanium alloys, Haynes 25 superalloy and Inconel 718 because of their high hot hardness, excellent corrosion resistance, and high wear resistance (Danish et al., 2022). The nickel-based superalloy Inconel 718 is widely used in the energy, aerospace, and aviation sectors for components that can tolerate high temperatures and high mechanical loads. However, the alloy's limited heat conductivity and high sensitivity to strain rate and strain complicate the machining process. The poor machinability results in a short tool lifespan and inferior surface quality. Furthermore, the low thermal conductivity of Inconel 718 results in high cutting temperatures in the machining areas (Le and Dudzinski, 2014; Li et al., 2002; Zhang et al. 2012; Musfirah et al. 2017; Sinha et al; 2023). The high temperature during machining changes the microstructure of the alloy, produces residual stresses on the surface and triggers tool wear mechanisms.

It is challenging to maintain good precision and surface smoothness when machining hard materials because of the significant heat generation that occurs during this process (Deiab et al., 2014; Gürbüz and Baday 2024). Several approaches have been examined to reduce cutting temperature, decrease tool wear, and enhance surface quality. Studies have analyzed the machining behavior of difficult-to-cut materials across various cutting environments. Fan et al. (2015) investigated the behaviour at the contact between the tool and chip during turning. Inconel 718 was machined by TiAlN tool. Tool life got enhanced under the moist air condition as compared to dry cutting condition. The oxides Cr_2O_3, Fe_2O_3, and Fe_3O_4 were generated during lubrication that prohibited the tool wear. Sarkar and Datta (2021) studied the turning characteristics of Inconel 718 using dry machining, MQL, and nanofluid-based MQL (NFMQL). Flank wear and tool tip temperature were assessed, showing greater wear with NFMQL than MQL above 83 m/min. Tool temperatures dropped by 60% with MQL and 67% with NFMQL compared to dry cutting.

Saleem and Mehmood (2022) conducted an experimental study on tool performance and surface integrity during dry and MQL turning of Inconel 718, using sunflower and castor oils as cutting fluids. The lowest CLA surface roughness (0.43 μm) and minimal tool wear (116 μm) were achieved with sunflower oil under MQL, representing reductions of 52% and 26%, respectively, compared to dry machining under identical parameters. Ross et al. (2022) examined the surface morphological properties in the milling of Nimonic 80A alloy in a hybrid (CO_2 + MQL) cooling technique that offered sufficient cooling and lubrication. The hybrid setup reduced the temperature and specific cutting energy by 34–53% and 17–19%, respectively, as compared to the MQL setup.

This chapter aims to investigate how various lubrication and cooling strategies influence the milling of Inconel 718, a difficult-to-machine alloy. The study compares three machining environments: dry cutting, compressed air cooling, and MQL using a bio-based cutting fluid derived from *Putranjiva roxburghii* oil. The effects of these cutting settings are assessed in terms of axial force, tool wear, surface polish, and thermal behavior. The primary contribution of this chapter is the implementation of a bio-based oil, specifically extracted from *Putranjiva roxburghii*, which was further formulated into nanofluids and an emulsion. The bio-emulsion was applied as a cutting fluid in the machining zone using the Minimum Quantity Lubrication (MQL) technique, while the nanofluid was directly applied to the workpiece prior to the commencement of machining. The lubrication and cooling effectiveness of these bio-derived fluids is evaluated against dry machining and air-assisted cooling conditions. Chapter 3 provides a detailed discussion on the methodology for preparing the nanofluids and emulsion, the experimental setup for dry machining and air cooling, and the application of bio oil-based emulsion through the MQL technique, along with the characterization methods used. The following is the format for the following sections of this chapter: The experimental details and technique used in the study are presented in Section 5.2. Section 5.3 describes the optimization technique used for process parameter selection. Section 5.4 explain the detailed discussion of the results obtained. Section 5.5 offers a comparative analysis of the machining performance under dry, air-cooling, bio-oil-based emulsion cooling with MQL conditions and nanofluids cooling. Finally, Section 5.6 summarizes the key findings of this chapter.

5.2 Experimental Details and Methodology

A vertical milling (Make: BFW Inder industries, Model: HF-2) system was used for the experimental trials available at central workshop, IIT Guwahati, India. Experiments were conducted for the face milling operation of Inconel 718 on a 5.5 kW conventional vertical milling machine. The workpiece selected for face milling was Inconel 718 superalloy. The Inconel 718 workpiece materials in the form of rectangular block was purchased from local market with dimensions of 125 mm × 100 mm × 13 mm. The machining process of the Inconel 718 was executed using carbide cutting tool inserts. A 20 mm diameter tool bit with cemented carbide milling cutter inserts (Widia PA 145) was utilized. These inserts featured a cemented carbide substrate with a coating of TiAlN.

The fluid used in MQL condition was an indigenously developed vegetable oil-based cutting fluid. For preparing the cutting fluid, Putranjiva roxbughii oil was used. The seeds of the Putranjiva roxburghii tree were collected from the campus of the National Institute of Technology Silchar, Assam, India. The seeds were then cleaned to remove dirt and ensure the oil was pure. A magnetic stirrer and emulsifying agent lecithin were used to prepare a proper emulsion. Six different emulsion samples were formulated by altering the volumetric ratio of water to oil (in milliliters) as follows: 100:5, 100:10, 100:15, 100:20, 100:25, and 100:30.

Machining was conducted in three different conditions, viz., dry, air cooling and vegetable oil-based MQL cooling. The machining variables were depth of cut (d_c), spindle speed (N) and feed rate (f); d_c was fixed at 0.1 mm for all cutting conditions. Face milling width was 13 mm and length 125 mm. Machining was conducted at six spindle rotations, viz. 45, 90, 180, 250, 355 and 500 revolutions per minute (RPM). Three different table feed rates were employed, viz., 16, 20 and 40 mm/min. The workpiece was face milled using two flute milling cutters with carbide tool inserts. For air cooling condition, an air compressor was utilized that facilitated compressed air supply at different pressure variations, viz., 0.5, 0.75, 1.0, 1.25 and 1.5 bar. This compressed airflow prevented the cutting tool from overheating. During the experiments, compressed air was continuously supplied by air flow regulator comprising a nozzle of diameter 3 mm to the machining area. Under MQL conditions, machining was carried out using cutting fluids derived from vegetable oil.

To evaluate the impact of nanofluids on machining performance, three formulations were selected from a total of twelve samples based on their superior thermal conductivity. These included 0.5 wt.% SiO_2, 0.75 wt.% graphene, and a hybrid nanofluid containing 0.25 wt.% SiO_2 + 0.25 wt.% graphene. The hybrid formulation was also chosen for its lowest contact angle, indicating enhanced wettability and improved lubrication at the tool–workpiece interface. For the experiments, 10 mL of the selected nanofluid was uniformly applied to the 120 mm × 13 mm face of a rectangular Inconel 718 block prior to face milling. The machining operation was performed over a length of 120 mm to assess the influence of the nanofluids on key performance parameters.

5.3 Optimization Technique

To determine the optimal set of machining input parameters, G.E.P. Box's evolutionary optimization technique as described in Deb (2012) was used. In this technique, for n decision variables, $2^n + 1$ points are considered to obtain the function values. Starting from an initially assumed best point x_0, a hypercube is constructed around this initial point comprising 2^n corner points. The function values at $2^n + 1$ points are evaluated and the best point is identified. The best point is the new x_0. In the next iteration, another n dimensional hypercube is created around x_0. The function values at all the corner points are evaluated and the best point is identified. This procedure is repeated until there is no change in the best point. Once two consecutive iterations provide the same best point, the size of the hypercube is reduced and a new hypercube is constructed. The whole process is repeated until the size of hypercube is sufficiently small.

In this study, table feed and spindle speed are considered as machining parameters. Three values of table feed (16, 20, 40 mm/min) and four spindle speeds (90, 180, 250 and 355 RPM) were considered. The first setting included a medium spindle speed and feed rate, or 20 mm/min and 250 RPM, respectively. In Section 5.4.1, the experimental trials' outcomes are covered in depth.

5.4 Results and Discussion

The experimental investigation was carried out in three separate phases. First, tests were performed under dry machining conditions without any external cooling. Next, experiments were executed using air cooling. Lastly, trials were conducted under MQL employing vegetable oil.

5.4.1 Performance of milling under dry condition

Dry machining experiments were carried out with systematic variation of feed rate and spindle speed as per Box's evolutionary optimization method illustrated in Figure 5.1. The first coordinate refers to the feed rate in mm/min and the second to the spindle speed in RPM. In iteration 1, the lower left corner has the coordinates (16, 180) and upper left corner has the coordinates (16, 355). The lower right corner has the coordinates (40, 180) and upper right corner has the coordinates (40, 355). The minimal surface roughness was measured at the combination (16, 180) after testing at all designated sites. This corresponds to a spindle speed of 180 RPM and a feed rate of 16 mm/min. This corresponds to 11.3 m/min of cutting speed and 0.044 mm of feed per tooth. Table 5.1 shows the results of iteration 1 of surface roughness, tool temperature and workpiece temperature.

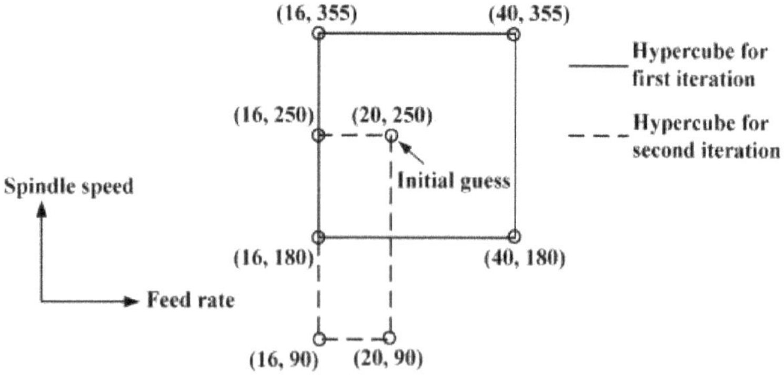

FIGURE 5.1 Design of experiments as per Box's evolutionary optimization method

In the second iteration, the coordinate (16, 180) was considered as the initial point. Another hypercube shown by the dotted line in Figure 5.1 was prepared, having a lower left coordinate (16, 90), upper left coordinate (16, 250), and lower right coordinate (20, 90). Due to machine limitation, the value of feed is considered 16 mm/min in both left upper and lower coordinates. It this iteration, the lowest surface roughness was obtained again at (16, 180).

TABLE 5.1 Results of Iteration 1

Exp. No.	Input Parameters			Experimental results		
	Depth of cut (mm)	**Feed rate (mm/min)**	**Spindle speed (RPM)**	**Roughness R_a (μm)**	**Tool temperature , T_t (°C)**	**Workpiece temperature , T_w (°C)**
1	0.1	20	250	1.19	61.63	47.05
2	0.1	40	355	1.93	65.03	48.83
3	0.1	16	180	0.64	52.57	40.9
4	0.1	16	355	1.12	61.7	47.14
5	0.1	40	180	1.56	59.46	43.88

Thus in both iterations 1 and 2, the obtained best point is the same. As per Box's algorithm, the hypercube should be reduced, but in this study, already the sufficient precision was achieved as per machine tool availability. No further reduction of hypercube was possible; hence, the obtained result was considered as the best result, and further study was carried out taking the feed rate of 16 mm/min and spindle speed of 180 RPM. The best surface roughness (R_a of 0.64 μm) was obtained at these values. Detailed discussion is presented in the following subsections. Table 5.2 shows the results of iteration 2 of surface roughness, tool temperature and workpiece temperature.

TABLE 5.2 Results of iteration 2

Exp. No.	Input Parameters			Experimental results		
	Depth of cut (mm)	**Feed rate (mm/min)**	**Spindle speed (RPM)**	**Roughness R_a (μm)**	**Tool temperature T_t(°C)**	**Workpiece temperature (°C)**
1	0.1	16	250	0.74	60.7	44.43
2	0.1	20	90	1.45	44.67	35.96
3	0.1	16	90	1.32	40.73	34.33

Study of surface roughness

The impact of tool speed on the workpiece's average surface roughness at different feed rates is shown in Figure 5.2. Surface roughness decreases as the spindle speed increases from 90 to 180 RPM. A feed rate of 16 mm/min at 180 RPM results in a surface roughness of 0.64 μm. This reduction in roughness up to 180 RPM is attributed to insufficient time for built-up edge formation. However, beyond 180 RPM, surface roughness increases due to heightened vibration and temperature. The maximum surface roughness value of 1.93 μm (Ra) was observed at the highest spindle speed of 355 RPM combined with a feed rate of 40 mm/min. Comparable findings were reported by D'Addona et al. (2017) for turning and Salur (2022) for milling of Inconel 718.

It is worthwhile to estimate ideal surface roughness values. The peak-to-valley surface value for the spindle speed of 180 RPM and feed rate 16 mm/min is estimated as follows. For two teeth cutter, feed per tooth is given as 16/ (180×2) = 0.044 mm/tooth. The nose radius of the tool was 0.5 mm. Hence, peak-to-valley height is (Lal G.K, 2013)

$$h = \frac{f_t^2}{8R} = \frac{0.044^2}{8 \times 0.5} = 4.84 \times 10^{-4} \text{ mm.} \tag{5.1}$$

Assuming triangular surface profile, the centre line average surface value is one fourth of this value, i.e., 0.12 μm, significantly lower than the experimental obtained minimum value of 0.64 μm.

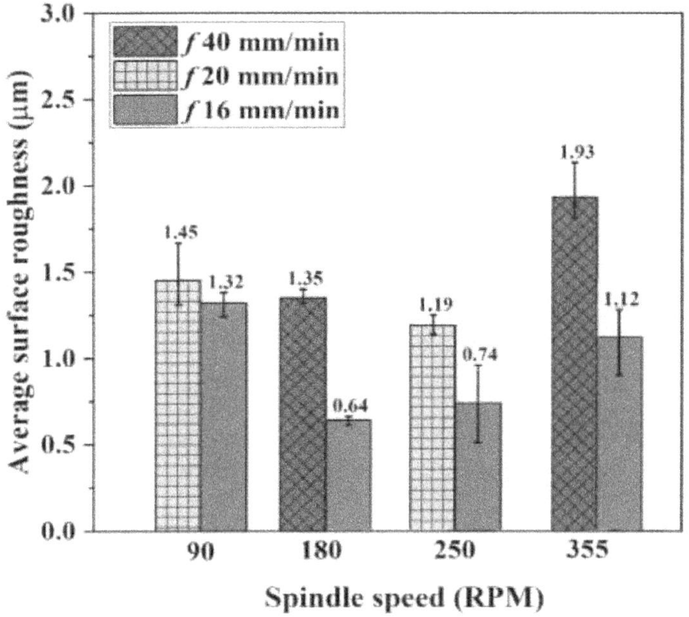

FIGURE 5.2 CLA surface roughness values for various combinations of spindle speeds and feed rates

Study of tool temperature

Variations in tool temperature during dry machining of Inconel 718 at various spindle speeds and feed rates are depicted in Figure 5.3. The tool temperature continuously increases at all feed rates as the spindle speed increases from 90 to 355 RPM. Tools have relatively low temperatures at 90 RPM spindle speed, ranging from 40.73 °C at a feed rate of 16 mm/min to 44.67 °C at a feed rate of 40 mm/min. The tool temperature increases as the spindle speed increases up to 180 RPM, reaching 59.46 °C at a feed rate of 40 mm/min. As spindle speed continues to rise, reaching a top of 355 RPM, the temperature trend continues to rise. At the 40 mm/min feed rate, the maximum temperature measured at this maximum speed was 65.03 °C. Additionally, a significant temperature of 61.7 °C was noted at the lower feed rate of 16 mm/min. At a spindle speed of 90 RPM and a feed rate of 16 mm/min, on the other hand, the lowest temperature recorded was 40.73 °C. The temperature decreased as the feed rate was reduced at all spindle speeds. The temperature of the tool increased noticeably when the spindle speed increased to 250 RPM. However, beyond 250 RPM and up to 355 RPM, the rate of temperature increase became less significant.

The increase in tool temperature with higher spindle speeds is due to high energy input in a given time; even the friction may increase due to temperature causing further increase in the temperature. Material removal rate (MRR) is also a factor for the enhancement of temperature at higher feed rates. Higher feed rates contribute further by increasing the MRR generating more heat. This increased material removal generates more heat. Also, at higher feed rates, the interaction between the tool and the workpiece becomes more significant. This leads to higher friction and consequently higher tool temperatures. A simulation-based study by Sulaiman et al. (2013) found that higher feed rates during machining lead to a rise in cutting temperature. Their study focused on the machining of titanium alloy.

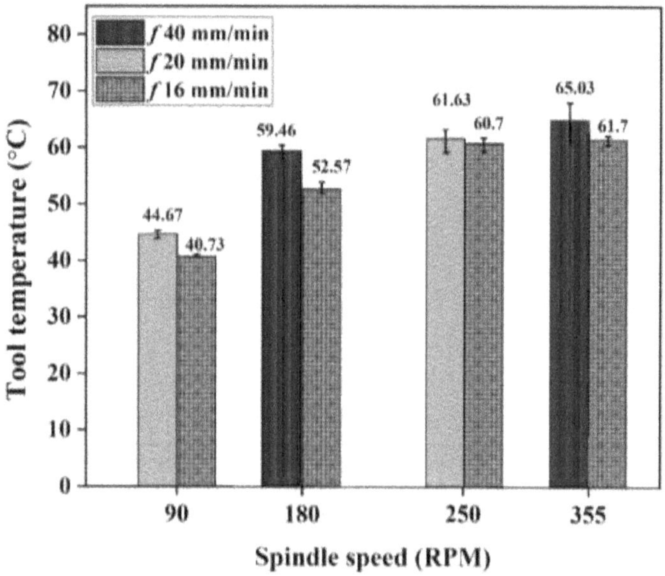

FIGURE 5.3 Variations of tool temperature with spindle speed in dry machining

Study of tool wear

Tool wear at different spindle speeds was analysed by SEM technique. At a speed of 90 RPM, abrasive wear and nose wear were observed as shown in Figure 5.4. It is due to asperities in the workpiece material rubbing against the tool surface causing material removal and scratches. At a speed of 180 RPM, there was a formation of built-up edge (BUE). BUE occurs when workpiece material welds onto the cutting edge during machining. Abrasive wear was observed. At a spindle speed of 250 RPM, there was a rise in the temperature up to 60.7 °C, causing oxidation of the tool material. There was also an indication of flank wear. The material of the workpiece and the tool's flank surface abrasively interact to cause flank wear.

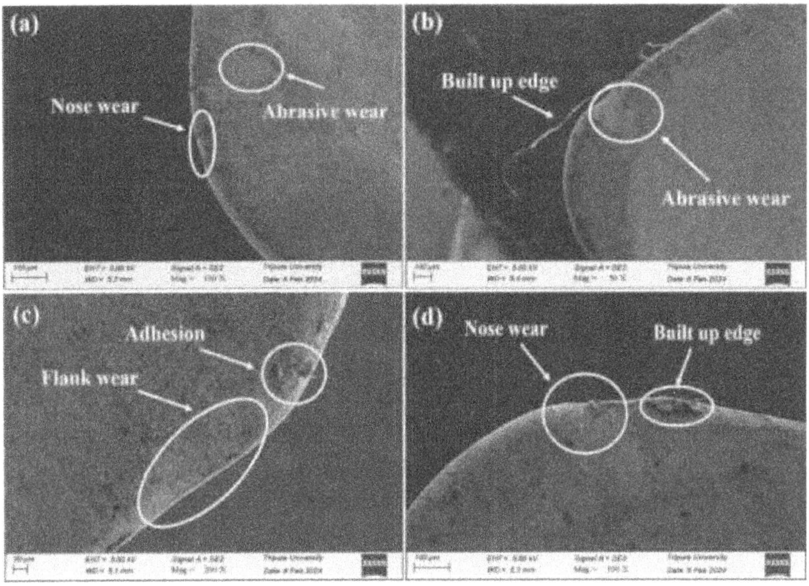

FIGURE 5.4 Tool wear at different spindle speeds (a) 90 RPM (b) 180 RPM (c) 250 RPM (d) 355 RPM in dry machining at a feed of 16 mm/min

Figure 5.5 presents the EDX analysis of the cutting tool operated at 180 RPM, revealing the presence of elements such as carbon (C), oxygen (O), chromium (Cr), nickel (Ni), and tungsten (W). The presence of W indicates the contribution of the tool substrate, which is primarily composed of tungsten carbide. Presence of oxygen suggests the influence of atmospheric condition during the milling process that occurred due to exposure to air. The presence of Cr and Ni are associated with the workpiece material. Their detection indicates the transfer of material during the machining process. It occurs due to tool-workpiece interaction.

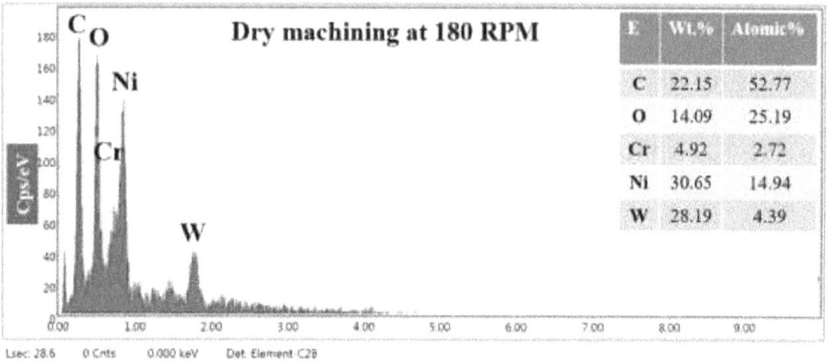

FIGURE 5.5 EDX analysis of tool at 180 RPM spindle in dry machining

5.4.2 Performance of milling in air cooling

For air cooling, spindle speeds of 90, 180, 250 and 355 RPM were considered for conducting the experiments. Compressed air from 0.5 to 1.5 bar pressure were considered. The feed rate was maintained at a constant value of 16 mm/min. In the next sections, specific topics like temperature fluctuations, tool wear, and surface roughness are covered in detail.

Study of surface roughness in air cooling environment

Figure 5.6 shows how surface roughness changes with spindle speed and varying air pressures of 0.5, 1.0, and 1.5 bar. At the minimum spindle speed of 90 RPM, the surface roughness is highest at 0.83 μm for 0.5 bar, followed by 0.75 μm at 1.0 bar and 0.72 μm at 1.5 bar. Increasing the spindle speed to 180 RPM resulted in a decrease in surface roughness for all air pressure levels, dropping to 0.54 μm, which suggests that elevated air pressure enhances the surface finish. Nonetheless, at 250 RPM, surface roughness rose for each pressure setting. Similarly, at the maximum spindle speed of 355 RPM, there was a further increase in surface roughness. This growth in roughness is attributed to elevated friction within the cutting zone.

The findings showed that higher air pressure helps decrease surface roughness, enhancing the machined surface quality. At 180 RPM and 1.5 bar of air pressure, the minimal surface roughness of 0.54 μm was attained. The maximum surface roughness, measured at 0.81 μm, occurred at 355 RPM with an air pressure of 0.5 bar.

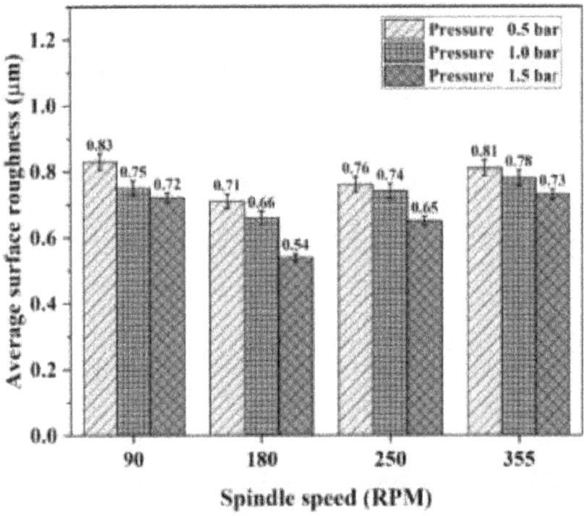

FIGURE 5.6 Variation of surface roughness with different spindle speeds and varying pressure at constant feed of 16 mm/min

To study how compressed air pressure influences surface roughness, tests were additionally performed at intermediate pressures of 0.75 and 1.25 bar, maintaining a spindle speed of 180 RPM and a feed rate of 16 mm/min. Surface roughness slightly increased from 0.71 μm at 0.50 bar to 0.76 μm at 0.75 bar as illustrated in Figure 5.7. However, at 1 bar, the roughness decreased to 0.66 μm. A slight increase in surface roughness is observed at 1.25 bar, where the roughness reaches 0.74 μm. The most significant reduction in surface roughness was at 1.5 bar, where it dropped to the lowest value of 0.54 μm.

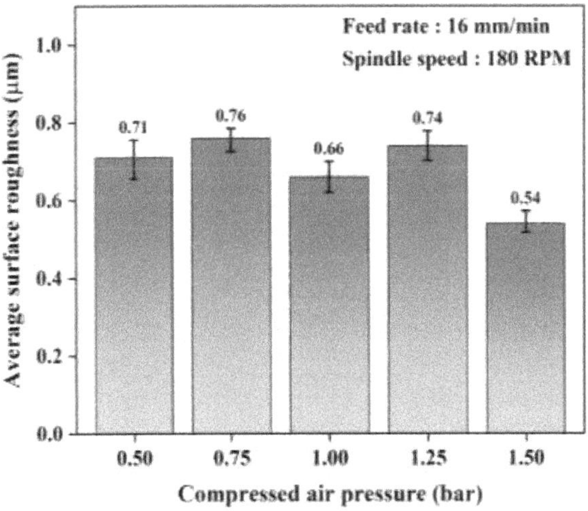

FIGURE 5.7 Variation of surface roughness with compressed air pressure

At higher pressures, the velocity of the air jet increases, which significantly improves convective heat removal from the cutting zone. Lower cutting temperatures reduce thermal softening of the workpiece material and minimize tool wear mechanisms such as adhesion, abrasion, and built-up edge formation (Mia and Dhar, 2015). Additionally, the high-velocity air stream enhances chip evacuation and prevents chip re-cutting, which otherwise damages the machined surface. These combined effects promote a more stable cutting action and maintain a sharper tool edge, resulting in improved surface finish.

Study of tool and workpiece temperaturein air cooling environment

Figure 5.8 illustrates the variation in tool and workpiece temperatures at different compressed air pressure, viz., 0.5, 0.75, 1, 1.25 and 1.5 bar. At a pressure of 0.5 bar, the tool surface temperature was 48.54 °C, while the workpiece temperature was 40.43 °C. As the pressure increased to 0.75 bar, both tool and workpiece temperatures slightly decreased, tool at 47.5°C and the workpiece at 37.43 °C. At 1 bar pressure, the tool temperature was 49.8 °C. This was the highest recorded temperature of the tool irrespective of all air pressures. The high temperature at 1 bar was due to insufficient cooling and increased frictional heat. When the pressure was raised to 1.25 bar, there was a reduction in both temperatures; tool temperature was 43.7 °C and the workpiece temperature was 35.85 °C. At the highest pressure of 1.5 bar, both tool and workpiece temperatures reached their lowest values of 42.5 °C and 34.93 °C, respectively. According to this, increased air pressure improves the tool's and the workpiece's cooling efficiency. It is to be noted that temperature reduction occurred at pressures beyond 1 bar.

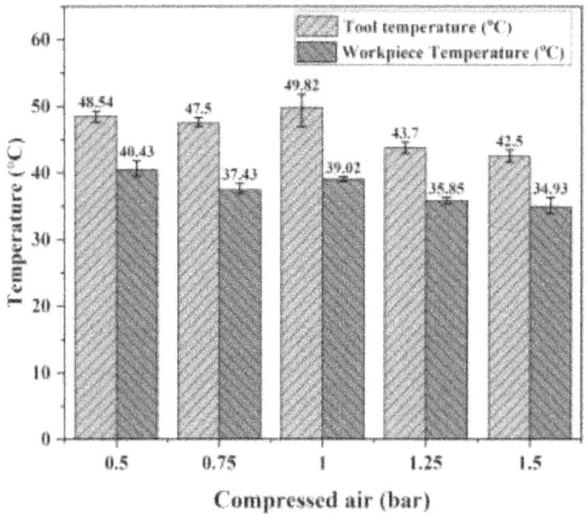

FIGURE 5.8 Tool and workpiece temperature variation with cutting velocity at different pressure

Study of tool wear in air cooling environment

FESEM images of tool wear at different air pressures, viz., 0.5, 1, 1.25 and 1.5 bar are illustrated in Figure 5.9. At a pressure of 0.5 bar, flank wear and BUE are observed in the tool surface, as shown in Figure 5.9 (a). Flank wear reflects intense friction occurring between the tool and the workpiece. The built-up edge (BUE) developed as a result of inadequate cooling, leading to elevated local temperatures that promote material sticking. In Figure 5.9 (b), notch wear and edge chipping are observed. This took place on the tool when subjected to a pressure of 1 bar. The notch wear appeared as a localized depression on the tool edge. Edge chipping indicated that the cutting edge of the tool was broken due to the combination of mechanical stress and inadequate cooling. The formation of BUE on the tool wear is also depicted in Figure 5.9 (c), i.e., at an air pressure of 1.25 bar. This indicates that the pressure was not sufficient to prevent material adhesion. However, the notch wear indicates ongoing stress on the tool due to tool vibration and localized heating. At the highest pressure of 1.5 bar, the presence of crater wear on the rake face of the tool indicates the occurrence of material erosion, as illustrated in Figure 5.9 (d).

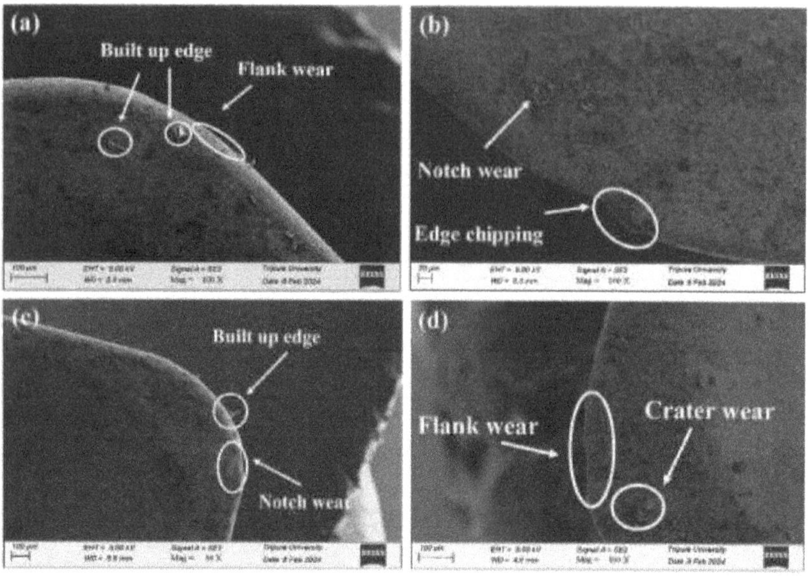

FIGURE 5.9 SEM images of tool wear at different air pressure in air cooling: (a) 0.5 bar, (b) 1 bar, (c) 1.25 bar, (d) 1.5 bar

Figure 5.10 shows the EDX analysis of tool at 180 RPM and an air pressure of 1.25 bar. The presence of 55.82 wt.% O indicated the significant oxidation that occurred during cooling process. The presence of Ni was associated with the workpiece material that was transferred to the tool during the machining process.

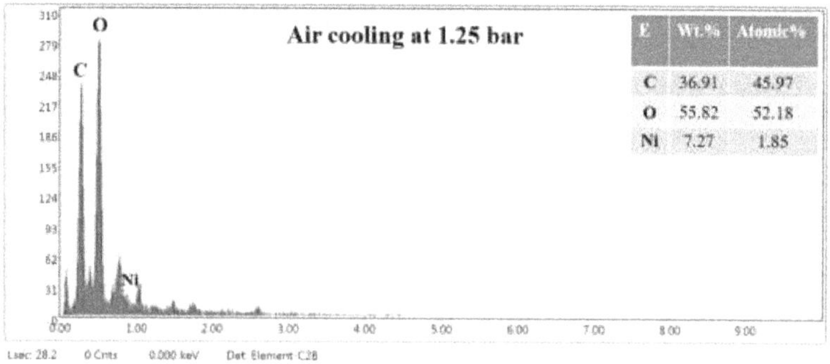

FIGURE 5.10 EDX analysis of tool at 180 RPM and air pressure of 1.25 bar

5.4.3 Performance of milling under vegetable oil based emulsion cooling

Putranjiva roxburghii vegetable oil was utilized as the base for preparing cutting fluid emulsions. Six emulsion samples were formulated by varying the water-to-oil volume ratios as follows: 100:5, 100:10, 100:15, and 100:20, 100:25, and 100:30 (ml). The thermal properties of these emulsions were evaluated prior to their application in the machining zone using the MQL technique. During machining, surface roughness was measured at four different spindle speeds 90, 180, 250, and 355 RPM while maintaining a constant feed rate of 16 mm/min. Each water-oil emulsion combination was applied individually in the machining zone under MQL conditions.

Study of thermal properties of emulsions

Figure 5.11 shows the variation of thermal conductivity of different emulsions. The lowest value of thermal conductivity, i.e., 0.558 W/m.K was obtained at the emulsion ratio of 100:5. As the concentration of the oil increased, the thermal conductivity of the emulsion also increased. The highest value of thermal conductivity was 0.574 W/m.K. It was obtained at the emulsion ratios of 100:20, 100:25 and 100:30. This indicated that increasing the concentration of oil after a certain point does not affect the thermal conductivity.

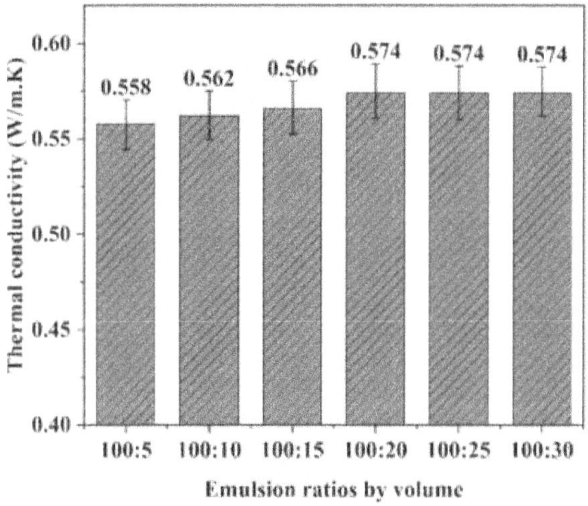

FIGURE 5.11 Variation of thermal conductivity at different emulsion ratios

The variation of volumetric heat capacity of different emulsions is shown in Figure 5.12. The volumetric heat capacity increased from 2.591 $MJ/m^3 \cdot K$ at the emulsion ratio of 100:5 to the highest value of 3.719 $MJ/m^3 \cdot K$ at an emulsion ratio of 100:20. However, a further increase in the emulsion ratio resulted in the reduction of volumetric heat capacity. This indicated that the optimal ratio for obtaining the highest value of volumetric heat capacity is 100:20, beyond which a reduction in the volumetric heat capacity was observed.

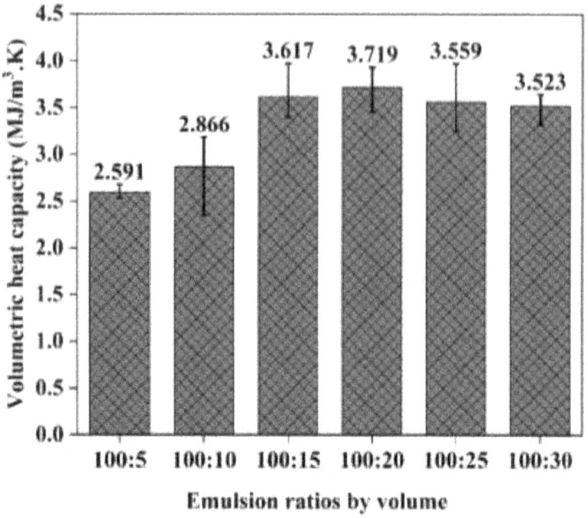

FIGURE 5.12 Variation of volumetric heat capacity at different emulsion ratios

Study of surface roughness in emulsion cooling

Figure 5.13 illustrates how surface roughness changes with varying emulsion ratios. The highest recorded surface roughness of 0.73 μm was obtained with an emulsion ratio of 100:5 at a spindle speed of 90 RPM. The surface roughness with the lowest value, 0.43 μm, was measured at 90 RPM and 100:20 emulsion ratio. It was observed that surface roughness decreased when the emulsion ratio was increased from 100:5 to 100:20. Nevertheless, surface roughness rose over this emulsion ratio value. At all emulsion ratio values, the surface roughness was reduced at 180 RPM spindle speed. At an emulsion ratio of 100:20, the lowest surface roughness value of 0.29 μm was attained. Nevertheless, increasing the spindle speed from 180 RPM to 355 RPM caused a rise in surface roughness across all emulsion ratios. The results suggest that both emulsion ratio and spindle speed influences the surface roughness. At higher concentration of vegetable oil, the cutting fluid become relatively more viscous. This causes the chips to stick to the workpiece and cutting tool resulting in a poor surface finish.

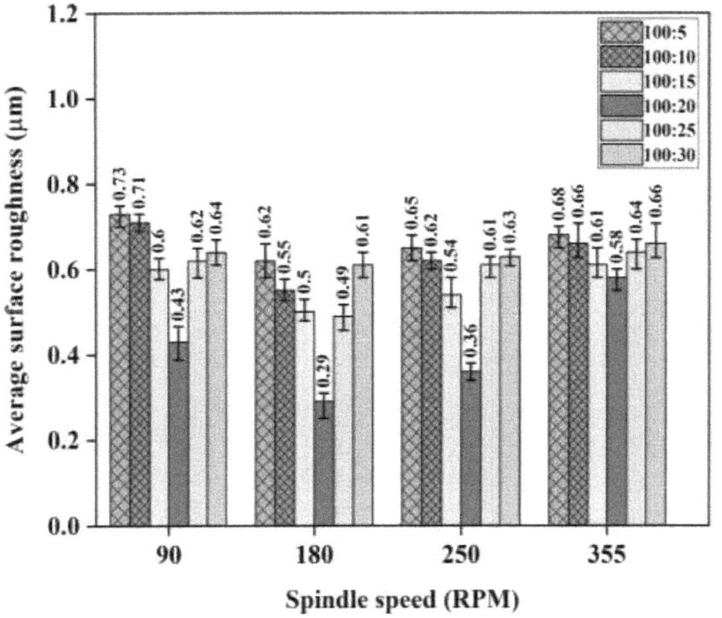

FIGURE 5.13 Variation of surface roughness with different spindle speed in different water-oil emulsions at constant feed rate 16 mm/min

Study of tool and workpiece temperature in emulsion cooling

Temperature variation during the MQL condition is shown in Figure 5.14. The spindle speed and feed rate in this investigation were kept at 180 RPM and 16 mm/min, respectively. It was found that raising the emulsion ratio led to a reduction in the temperatures of both the tool and the workpiece. At the emulsion ratio of 100:20, the lowest temperature was achieved for the tool (41.1 °C) as well as workpiece (31.46 °C) surfaces. However, with a further increase in emulsion ratio from 100:20 to 100:30, both tool and workpiece surface temperature increased. There is a noticeable difference in the tool's temperature compared to the workpiece. Based on the experimental results, the 100:20 emulsion ratio was found to be the most effective. This ratio provided the lowest surface roughness due to improved lubrication at the tool–workpiece interface, which minimized frictional effects and reduced ploughing. At the same time, the enhanced thermal conductivity of the emulsion at this ratio facilitated better heat dissipation, resulting in lower cutting-zone temperatures.

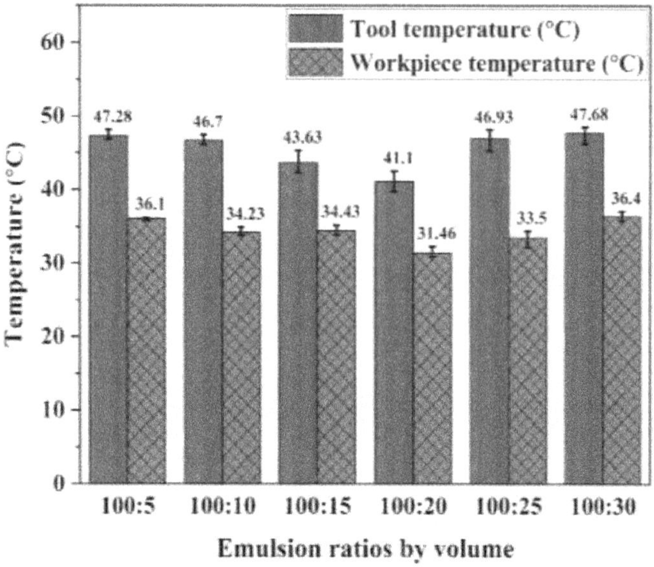

FIGURE 5.14 Variation of tool and workpiece temperatures with different emulsions ratios at spindle speed of 180 rpm and feed rate 16 mm/min

Study of tool wear in emulsion cooling

Figure 5.15 depicts the tool wear observed at various emulsion ratios. The FESEM image in Figure 5.15 (a) displays the tool wear at an emulsion ratio of 100:5, where crater wear occurred due to elevated temperatures at the tool-chip interface. Flank wear was detected at the emulsion ratio of 100:10, as shown in Figure 5.15 (b). This is due to the continuous abrasion with the workpiece material leading to gradual material removal from the flank face. At the emulsion ratio of 100:15, notch wear was observed due to localized wear at the cutting edge as shown in Figure 5.15 (c). At the emulsion ratio of 100:20, crater wear was observed as shown in Figure 5.15 (d). As the emulsion ratio increases up to 100:25, notch wear, BUE and flank wear were observed due to adhesion and abrasion with the workpiece material. At a higher emulsion ratio of 100:30, flank wear was noted due to abrasive action between the tool and the workpiece material.

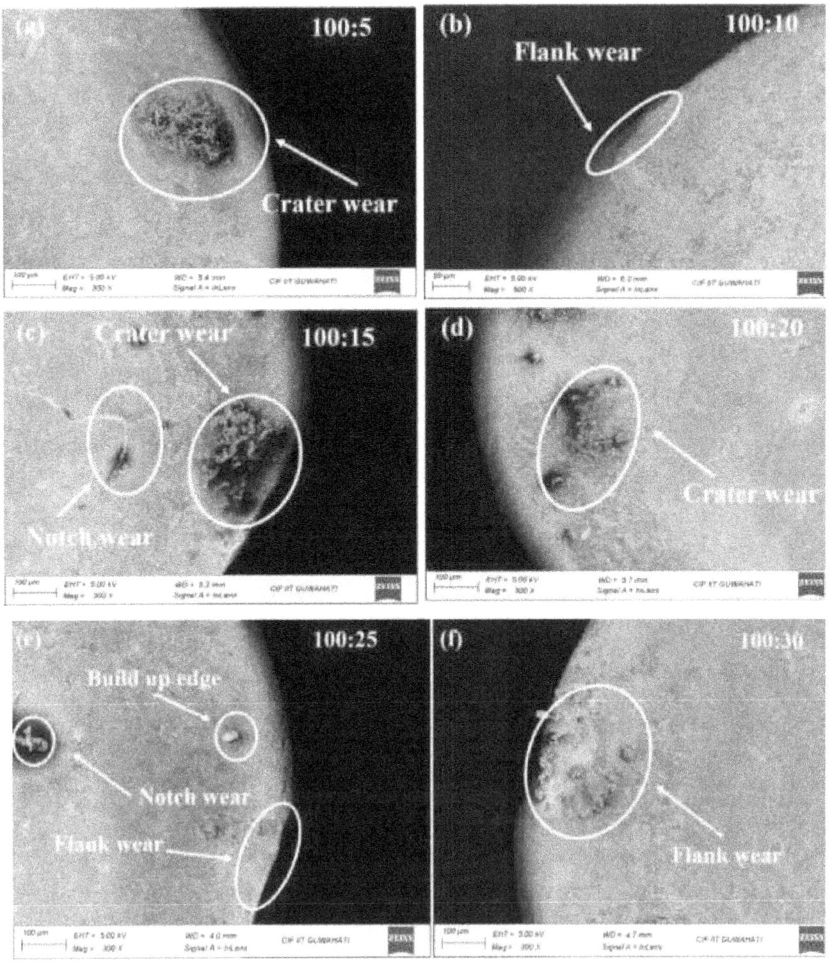

FIGURE 5.15 SEM images of tool wear in different emulsion ratios: (a) 100:5, (b) 100:10, (c) 100:15, (d) 100:20, (e) 100:25, (f) 100:30

5.4.4 Performance of milling under vegetable oil based nanofluid cooling

In this study, first, experiments were conducted in dry and base oil wetted conditions. Further experiments were conducted using various nanofluids, viz., SiO_2-based, graphene-based and hybrid nanofluids. Weight percentages of SiO_2 and graphene in Putranjiva roxburghii were 0.5 and 0.75 for SiO_2-based and graphene-based nanofluids, respectively. The hybrid nanofluid contained 0.25 wt.% SiO_2 and 0.25 wt.% graphene nanoparticles in Putranjiva roxburghii. These compositions provided the lowest contact angles, indicating high wettability. Following several trial runs, the machining parameters were finalized as a depth of cut of 0.1 mm, a spindle speed of 180 revolutions per minute (RPM), and a table feed rate of 16 mm/min. These settings correspond to a surface cutting speed of 11.3 meters per minute and a feed rate of 0.044 mm per tooth. These parameters were chosen based on an earlier study by (sharma et al. 2024). The length of the chip l can be evaluated by (Lal, 2013).

$$l = B\left[1+\left(\frac{B}{2D}\right)^2\right]^{1/2}, \tag{5.2}$$

Where B is the width of the workpiece and D is the diameter of the cutter. The average undeformed chip thickness t_{av} is evaluated as

$$t_{av} = \frac{v}{nN}\left[1-\frac{1}{2}\left(\frac{B}{2D}\right)^2\right], \tag{5.3}$$

Where v is the table speed, n is the number of teeth and N is the speed of the cutter in rpm. For B = 13 mm, D = 20 mm, v = 11.3 m/min, n = 2 and N = 180 rpm, the values of l and t_{av} are obtained as 13.67 mm and 0.03 mm, respectively. Taking these parameters as reference, additional experiments were carried out at spindle speed of 90 rpm and 250 rpm; both cutting speed and chip thickness get changed in proportion to spindle speed.

Study of surface roughness in nanofluids cooling

Figure 5.16 presents the average surface roughness of the workpiece produced under various machining conditions. The highest value of centreline average (CLA) surface roughness (1.29 μm) was obtained at dry machining at 90 rpm. The CLA surface roughness at 180 rpm and 250 rpm were 1.06 μm and 1.18 μm, respectively. Ideally, the surface roughness should reduce with speed. Nevertheless, higher speeds amplify vibrations, which lead to a rise in surface roughness. Hence, the lowest surface roughness was obtained at 180 rpm. The application of base oil reduced the surface roughness to 1.15 μm, 0.95 μm and 0.98 μm at 90 rpm, 180 rpm and 250 rpm, respectively. The lubrication provided by the oil assisted in smoother chip removal and reduced the tool-workpiece friction, leading to a better surface finish. The addition of 0.5 wt.% SiO_2 nanoparticles further reduced the roughness to 1.07 μm, 0.88 μm and 0.93 μm at 90 rpm, 180 rpm and 250 rpm, respectively. The nanoparticles enhanced the tribological properties of the base oil resulting in a finer finish. The CLA surface roughness obtained using graphene nanofluid was 0.91 μm at 180 rpm. The hybrid nanofluid achieved the minimum surface roughness of 0.86 μm at 180 RPM, attributed to its superior wettability characterized by a contact angle of 23.65°. Regardless of the machining environment, the lowest surface roughness consistently occurred at 180 RPM. These variations in surface roughness demonstrate that adding nanoparticles to the base oil can improve the workpiece's surface finish.

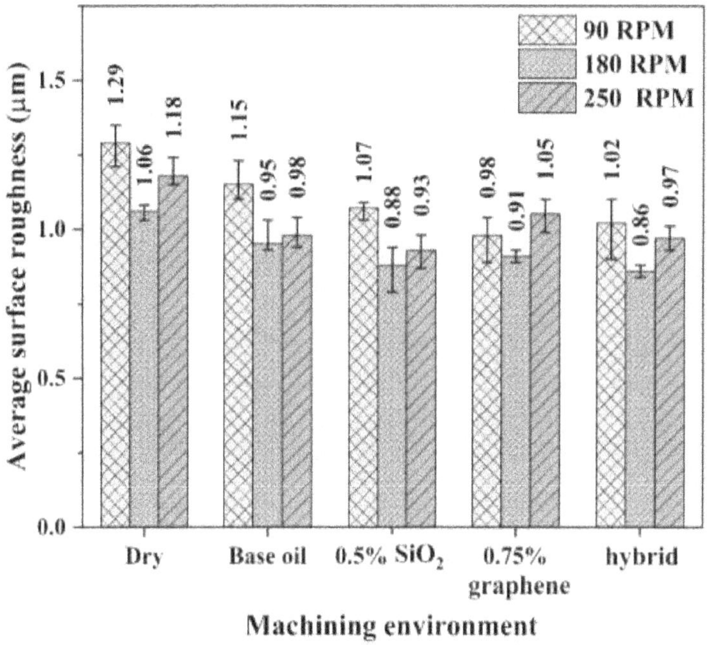

FIGURE 5.16 Variation of surface roughness in different machining environments

Study of tool temperature in nanofluids cooling

Figures 5.17 and 5.18 show the temperature variations of cutting tool and workpiece temperature variations of workpiece in different machining environments, respectively. It is to be noted that the measured temperatures are not the temperatures at the tool-work interface, which will be much higher. These temperature values represent the average measurements taken near the interface area using an infrared thermal camera. The exact temperature at the tool-work interface needs sophisticated measuring techniques or inverse estimation through some simulation tool. However, the measurements taken at the same spot in this study give a comparative idea about different cutting environments. In dry machining, the average (as measured by infrared thermal analyzer) tool and the workpiece temperatures are the highest, i.e., 62.8 °C and 47.2 °C, respectively. These values were obtained at a spindle speed of 250 rpm. It is due to the absence of any cooling effect resulting in maximum heat generation. The introduction of base oil reduced the tool temperature to 44.2 °C at 90 rpm, 55.6 °C at 180 rpm and 56.5 °C at 250 rpm as shown in Figure 5.17.

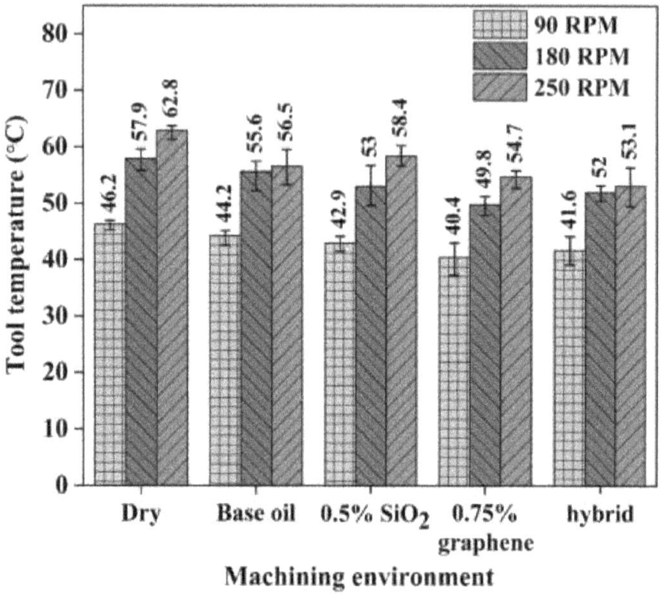

FIGURE 5.17 Variation of tool temperature in different machining environments

Study of workpiece temperature in nanofluid cooling

The workpiece temperature was reduced to 34.3 °C at 90 rpm, 43.1 °C at 180 rpm and 44.7 °C at 250 rpm as shown in Figure 5.18. This reduction in temperature is due to the lubricating and cooling properties of the oil. Adding 0.5 wt.% SiO_2 nanoparticles to the base oil further reduced the tool and workpiece temperature to 53 °C and 42.1 °C (at 180 rpm), respectively. SiO_2 nanoparticles enhanced the thermal conductivity and heat absorption capacity of the base oil thereby improving its cooling ability. The lowest values of tool and workpiece temperatures were obtained for cutting fluid comprising 0.75 wt. % graphene nanoparticles. At 90 rpm, the tool temperature was reduced to 40.4 °C and workpiece temperature to 31.2 °C. In hybrid (0.25 wt.% SiO_2 and 0.25 wt.%) nanofluid, the tool temperature at 180 rpm is higher than that of temperature obtained with graphene nanofluid, but lower than that of temperature obtained with SiO_2 nanofluid. However, the workpiece temperature obtained using hybrid nanofluid is higher than that the temperature obtained with SiO_2 as well as graphene nanofluids. This proves that nanofluids based on Putranjiva roxburghii are efficient cutting fluids.

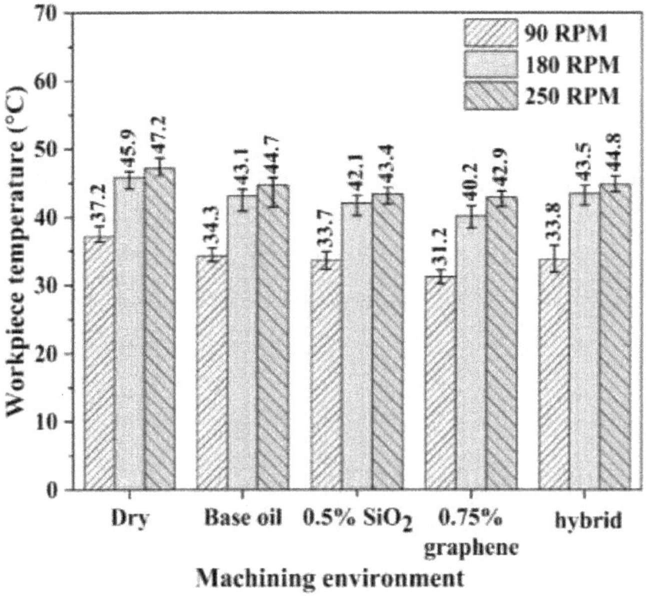

FIGURE 5.18 Variation of workpiece temperature in different machining environments

Study of chip morphology in nanofluids cooling

Figure 5.19 displays the chips formed under various machining conditions. These chips were produced using a depth of cut of 0.1 mm, a spindle speed of 180 RPM, and a table feed rate of 16 mm/min. In dry machining, wide chips were obtained with a maximum width of 1.24 mm as shown in Figure 5.19 (a). Machining without coolant led to increased friction and cutting forces at the tool-chip contact area, causing greater plastic deformation and the formation of thicker chips. Application of base oil resulted in the reduction of chip width up to 0.74 mm as shown in Figure 5.19 (b). Base oil reduced the temperature in the cutting zone and resulted in reduced plastic deformation. Figure 5.19 (c) illustrates the chip morphology obtained by machining with the addition of 0.5 wt.% SiO_2 nanoparticles to the base fluid.

The lubrication and heat dissipation properties of the cutting fluid were enhanced. As a result, chip width got reduced. Application of graphene based nanofluid further reduced the chip width up to 0.45 mm, as shown in Figure 5.19 (d). Chips obtained were the smallest among all machining environments. For hybrid based nanofluid, chip width was 0.68 mm. Chip lengths and thicknesses were highly non-uniform with frequent chip breakage. Due to small size and non-uniform thickness, proper thickness measurement could not be carried out. However, the thickness measurement at a few spots in a microscope provide some idea about chip deformation. As mentioned earlier, theoretical average undeformed thickness was 0.03 mm.

In dry machining, the typical chip thickness was 0.1 mm indicating huge deformation. In base oil and graphene based nanofluid, the average chip thickness was about 0.03 mm indicating low deformation in thickness direction; the deformation in width direction was still significant but much lower than that in dry machining. Variation in chip morphology in different cutting environments can be an indicator of machining performance.

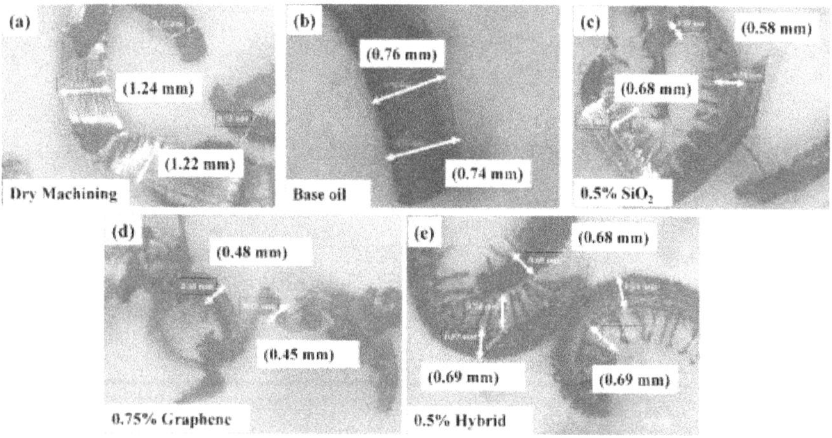

FIGURE 5.19 Chip morphology obtained under different machining environments: (a) dry machining, (b) base oil, (c) 0.5 wt.% SiO_2 based nanofluid, (d) 0.75 wt. % graphene based nanofluid, and (e) hybrid (0.25 wt.% SiO_2 and 0.25 wt.% graphene) nanofluid.

5.5 Comparative Analysis of Dry, Air-cooling and Bio-oil Based Minimum Quantity Lubrication

This experimental investigation focused on estimating the surface roughness and temperatures of both the tool and workpiece in dry, air-cooling and bio-oil based minimum quantity lubrication in the milling of Inconel 718 alloy. The lowest surface roughness values were 0.64 μm, 0.54 μm and 0.29 μm for dry, air-cooling and bio-oil based MQL environments, respectively. MQL achieved a 54.7% reduction in surface roughness as compared to dry conditions and a 46.3 % reduction as compared to air cooling. Air cooling achieved only a 15.6 % reduction in surface roughness as compared to dry condition.

In dry machining, the tool and workpiece temperatures reached 52.6 °C and 40.9 °C, respectively. Air cooling reduced these temperatures to 42.5 °C for the tool and 34.9 °C for the workpiece, indicating a 19.1 % reduction in tool temperature and a 14.6 % reduction in workpiece temperature. Under MQL conditions, the temperatures dropped further to 41.1 °C for the tool and 31.5 °C for the workpiece. The tool temperature was reduced by 21.8 % and workpiece temperature 23.1 % reduction as compared to dry machining.

5.6 Summary

This chapter examines the machining behavior of Inconel 718 superalloy across three different cooling methods: dry machining, air cooling, and MQL utilizing Putranjiva roxburghii oil-based fluids. Air cooling was conducted at five distinct pressure settings: 0.5, 0.75, 1.0, 1.25, and 1.5 bar. For the MQL approach, six emulsion mixtures were created by adjusting the water-to-oil volume ratios (in mL) to 100:5, 100:10, 100:15, 100:20, 100:25, and 100:30. For evaluating machining performance with nanofluids, three formulations with the highest thermal conductivity were selected from a total of twelve samples. These included 0.5 wt.% SiO_2, 0.75 wt.% graphene, and a hybrid formulation containing 0.25 wt.% SiO_2 + 0.25 wt.% graphene. Axial force, tool temperature, tool wear, and surface roughness were the criteria used to assess machining performance. The following is a summary of the experimental study's findings.

- The best value of surface roughness was obtained as 0.29 μm in MQL conditions. This value is 54.7 % less as compared to the surface roughness in dry conditions and 46.3 % less as compared to air cooling. The substantial improvement in the surface roughness demonstrated the superiority of Putranjiva roxburghii oil.
- The lowest temperature of the tool and workpiece was obtained at MQL condition at an emulsion ratio of 100:20. The tool temperature is 21.8 % less and workpiece temperature is 23.1 % less as compared to the temperatures obtained in dry machining.
- Microstructural study analysed the tool wear under different machining condition. The tool wear identified in dry machining conditions were abrasive flank wear and significant formation of BUE. The wear reduced under MQL conditions, suggesting improved tool longevity.

- Adding nanoparticles to the base oil lowered the surface roughness from 1.06 μm during dry machining to 0.86 μm when using the hybrid nanofluid at a spindle speed of 180 RPM. Additionally, the tool temperature dropped from 57.9°C under dry conditions to 49.8°C with the application of graphene nanofluid. These findings highlight the enhanced effectiveness of nanoparticle-enriched fluids in optimizing machining performance.

The next chapter explores the impact of Putranjiva roxburghii oil-based emulsions enhanced with nanoparticles referred to as nanoemulsions on the machining performance of the difficult-to-machine material Inconel 718. The study focuses on evaluating how the incorporation of nanoparticles into the bio-based emulsion influences critical machining responses such as surface finish, tool wear, cutting temperature, and cutting forces. The results are compared against conventional dry, air-cooled, and base bio-emulsion conditions to highlight the effectiveness of nanoemulsions in sustainable and high-performance machining of superalloys.

References

D'addona DM, Raykar SJ, Narke, MM, et al. High speed machining of Inconel 718: tool wear and surface roughness analysis. *Procedia CIRP 2017*; 62: 269–274.

Danish M, Aslantas K, Hascelik A, et al. An experimental investigation on effects of cooling/lubrication conditions in micro milling of additively manufactured Inconel 718. *Tribology International* 2022; 173: 107620.

Deb K. Optimization for Engineering Design: Algorithms and Examples. PHI Learning Private, Limited, New Delhi, India 2012.

Deiab I, Raza S W, Pervaiz S. Analysis of lubrication strategies for sustainable machining during turning of titanium Ti 6Al 4V alloy. *Procedia CIRP* 2014; 17: 766 771.

Fan Y, Hao Z, Lin J, et al. New observations on tool wear mechanism in machining Inconel 718 under water vapour + air cooling lubrication cutting conditions. *Journal of Cleaner Production* 2015; 90: 381–387.

Gürbüz H and Baday Ş. Comprehensive investigation of the effect of cryogenic process on machining of Inconel 718 superalloys with uncoated end mills. *Proceedings of the Institution of Mechanical Engineers, Part E: Journal of Process Mechanical Engineering* 2024; 09544089241256758.

Le Coz G and Dudzinski D. Temperature variation in the workpiece and in the cutting tool when dry milling Inconel 718. *The International Journal of Advanced Manufacturing Technology* 2014; 74: 1133–1139.

Li L, He N, Wang M, et al. High speed cutting of Inconel 718 with coated carbide and ceramic inserts. *Journal of Materials Processing Technology* 2002; 129(1–3): 127–130.

Mia, M. and Dhar, N.R., 2015. Effect of high pressure coolant jet on cutting temperature, tool wear and surface finish in turning hardened (HRC 48) steel. *Journal of Mechanical Engineering*, *45*(1), pp.1-6.

Musfirah AH, Ghani JA, Haron CC. Tool wear and surface integrity of inconel 718 in dry and cryogenic coolant at high cutting speed. *Wear* 2017; 376: 125–133.

Obratańsk-Karmiris P, Karkalos NE, Kudelski R, et al. Experimental study on the effect of the cooling method on surface topography and workpiece integrity during trochoidal end milling of Incoloy 800. *Tribology International* 2022; 176: 107899.

Ross NS, Gopinath C, Nagarajan S, et al. Impact of hybrid cooling approach on milling and surface morphological characteristics of Nimonic 80A alloy. *Journal of Manufacturing Processes* 2022; 73: 428–439.

Saleem MQ and Mehmood A. Eco-friendly precision turning of superalloy Inconel 718 using MQL based vegetable oils: tool wear and surface integrity evaluation. *Journal of Manufacturing Processes* 2022; 73: 112–127.

Salur E. Understandings the tribological mechanism of Inconel 718 alloy machined under different cooling/lubrication conditions. *Tribology International* 2022; 174: 107677.

Sarkar S and Datta S. Machining performance of Inconel 718 under dry, MQL, and nanofluid MQL conditions: application of coconut oil (base fluid) and multi-walled carbon nanotubes as additives. *Arabian Journal for Science and Engineering* 2021; 46: 2371–2395.

Sharma, D.K., Singh, P.K., Sharma, F. and Dixit, U.S., 2025. Development and characterization of Putranjiva roxburghii with addition of SiO2 and graphene nanoparticles for using it as a cutting fluid. *Proceedings of the Institution of Mechanical Engineers, Part C: Journal of Mechanical Engineering Science*, *239*(11), pp.4162-4181.

Sinha MK, Kishore K, Sharma P. Surface integrity evaluation in ecological nanofluids assisted grinding of Inconel 718 superalloy. *Proceedings of the Institution of Mechanical Engineers, Part E: Journal of Process Mechanical Engineering* 2023; 09544089231171042.

Sulaiman S, Roshan A, Borazjani S. Effect of cutting parameters on cutting temperature of TiAL6V4 alloy. *Applied Mechanics and Materials* 2013; 392: 68–72.

Virdi RL, Chatha SS, Singh H. Processing characteristics of different vegetable oil-based nanofluid MQL for grinding of Ni-Cr alloy. *Advances in Materials and Processing Technologies* 2022; 8(1): 210–223.

Zhang S, Li JF, Wang, YW. Tool life and cutting forces in end milling Inconel 718 under dry and minimum quantity cooling lubrication cutting conditions. *Journal of Cleaner Production* 2012; 32: 81–87

CHAPTER

6

Performance of Putranjiva Roxburghii Oil Based Nanoemulsions in Milling of Inconel 718

6.1 Introduction

With the growing demand for high-performance materials in engineering applications, industries are increasingly utilizing hard-to-machine alloys of nickel, cobalt, titanium and iron. Inconel 718, a nickel-based superalloy, is widely used in the aerospace, marine, and nuclear industries for parts that endure severe mechanical and thermal stresses (Mahesh et al. 2021). However, the poor machinability of Inconel 718 poses significant challenges during machining operations. Its lower thermal conductivity, high strength, high hardness and work hardening tendency lead to high temperature at the tool–workpiece interface, which accelerates tool wear and deteriorates surface quality as well as dimensional accuracy.

Researchers investigated the influence of cutting parameters on the machinability of Inconel 718. Thakur et al. (2009) studied its machinability during turning with a tungsten carbide insert under dry cutting conditions. The cutting speed and feed were varied in the range of 40–60 m/min and 0.05–0.09 mm/rev, respectively, while the depth of cut was kept constant at 0.5 mm. Surface roughness was found to decrease with increasing cutting speed up to 55 m/min; however, no further improvement was noticed beyond this point. In a study by Rakesh and Datta (2019) assessed Inconel 718's dry turning performance with an uncoated carbide tool. The depth of cut and feed rate stayed at 0.4 mm and 0.1 mm/rev, respectively, while the cutting speed was set at 50, 75, 100, and 125 m/min. According to the results, cutting force decreased as cutting speed increased from 50 to 75 m/min; nevertheless, at higher speeds, force increased as a result of rapid tool deterioration.

Conventional petroleum-based oils pose environmental hazards, disposal challenges and health risks to operators (Roushan et al. 2024). Evaluating the performance, cost-effectiveness, and environmental impact of a cutting fluid is essential for sustainable machining (Dixit et al. 2012). Researchers are exploring vegetable oil-based cutting fluids such as sunflower, rapeseed, coconut and palm oils as sustainable alternatives to conventional lubricants. Mishra et al. (2024) investigated the finish milling process of Inconel 718 produced through the additive manufacturing technique known as powder bed fusion. Experiments were conducted with aluminium rich coated tool under dry, flood, MQL and Cryo-MQL environments. In the MQL setup, sunflower oil was used. Cryo-MQL improved machinability by reducing milling forces by 47.2% and flank wear by up to 67%, Compared to dry and wet conditions, Cryo-MQL provided lower surface roughness. Dry milling caused severe damage, including built-up edge formation and cutting-edge fracture. The Cryo-MQL technique led to smoother chip formation, reduced deformation and lower interface friction.

Recent studies have emphasized the improvement of vegetable oil-based cutting fluids by adding nanoparticles to boost their effectiveness in machining operations. Shokrani et al. (2022) conducted an investigation on the end milling of Inconel 718 using a MQL system, where rapeseed oil infused with electrically charged tungsten disulfide (WS_2) nanoparticles was used. During the experiment, the cutting speed, feed rate, and depth of cut were maintained at fixed values. Different cooling conditions, including flood cooling, MQL with oil (with and without electrical charge) and MQL with oil containing nanoparticles (with and without electrical charge) were evaluated. The electrically charged WS_2 nanofluid improved tool life by 10% and 90% as compared to flood cooling and conventional MQL, respectively.

In line with aforesaid studies on sustainable machining, this study examines the milling performance of Inconel 718 using *Putranjiva roxburghii* oil as a cutting fluid. Due to importance and complex nature of milling, it needs thorough investigation as highlighted by (Bartolomeis et al. 2021). The uniqueness of this study lies in employing Putranjiva roxburghii oil, a rarely investigated non-edible vegetable oil, as a cutting fluid for machining applications. This oil is derived from the Putranjiva roxburghii (also known as Drypetes roxburghii) tree of the Euphorbiaceae family found in tropical parts of India. This oil offers an economical and environmentally friendly alternative to conventional cutting fluids. Also, its medicinal properties and skin-friendly nature make it a safer option for machine operators (Supriya et al. 2017).

Recently, Sharma et al. (2025) examined the lubrication potential of Putranjiva roxburghii oil, developing nanofluids by dispersing varying concentrations of SiO_2 and graphene nanoparticles into the base oil. In a related study, Sharma et al. (2024) evaluated its performance in the milling of Inconel 718 using different water-to-oil emulsion ratios, but without incorporating nanoparticles. Addressing this research gap, this work evaluates the machining performance of *Putranjiva roxburghii* oil as a base in nano-emulsion, impregnated with SiO_2 and graphene nanoparticles. This investigation contributes to Goal 3 (Good Health and Well-being), Goal 12 (Responsible Consumption and Production), and Goal 13 (Climate Action) of United Nations Sustainable Development Goals by promoting cleaner technologies, minimizing occupational health risks and utilizing renewable resources (*Ref of* United Nations goals, Accessed on May 15, 2025)

6.2 Experimental Details

Nanoemulsions were prepared by combining water, Putranjiva roxburghii base oil and nanoparticles. Sharma et al. (2024) formulated several emulsions by varying the water-to-Putranjiva roxburghii oil volume ratio as 100:5, 100:10, 100:15, 100:20, 100:25 and 100:30 to investigate milling of Inconel 718. They reported that the emulsion comprising 100 ml water and 20 ml Putranjiva roxburghii oil provided the most favourable machining performance. In a related investigation, Sharma et al. (2025) evaluated the lubricating performance of Putranjiva oil with the addition of varying weight percentages of SiO_2 and graphene nanoparticles. Hybrid nanofluids of SiO_2 and graphene nanoparticles were also prepared by combining them in equal proportions. The highest thermal conductivity was achieved by 0.5 wt.% SiO_2, 0.75 wt.% graphene and 1 wt.% hybrid nanoparticles.

Based on previous study, three distinct nanoemulsions were prepared for use in MQL based cooling in milling of Inconel 718. The first nanoemulsion comprised 100 ml of water and

20 ml of Putranjiva roxburghii base oil with the addition of 0.5 wt.% SiO_2 nanoparticles. The second nanoemulsion was the base emulsion (100:20) incorporated with 0.75 wt.% graphene nanoparticles. The third was a hybrid nanoemulsion containing 0.5 wt.% of SiO_2 and 0.5 wt. % graphene nanoparticles in the base emulsion (100:20). Nanoemulsions were prepared as per the procedure outlined by Sharma et al. (2025), utilizing a magnetic stirrer (Make: REMI, Model: 2-MLH) and a probe sonicator (Make: SONICS Vibra cell, Model: VCX750). The thermal conductivity and specific heat of the prepared nanoemulsions were evaluated using a thermal properties analyzer (Make: Decagon Devices, Model: KD2 Pro). The contact angle of the prepared nanoemulsion was determined using a Goniometer (Make: Holmarc, Model: HO-IADCAM-01B).

To access the effectiveness of the nanoemulsions in MQL based cooling, experiments were conducted for face milling of Inconel 718 in a 5.5 kW conventional vertical milling machine (Make: BFW, Model: HF-2). Machining was done by cemented carbide milling cutter inserts (Widia PA 145). Further details of experimental setup, workpiece and tool are available in the work of Sharma et al. (2024) and are therefore not repeated here. Drawing from earlier results, the selected machining conditions for this study included a spindle speed of 180 RPM, a depth of cut of 0.1 mm, and a feed rate of 16 mm/min, identified as optimal parameters. Tool and workpiece temperatures were measured by thermal infrared measuring instrument (Make: Decagon Devices, Model: KD2 Pro) and surface roughness by Pocket Surf instrument (Make: Mahr Federal, Model: PMD-90101). The surface features of the cutting tools were examined through field emission scanning electron microscopy (FESEM) (Make: Zeiss, Model: Sigma 300).

To measure the cutting force, a 20 HP friction stir welding machine (Make: Royal Machine Tools, Model: RMV-1800) was used for face milling. Axial force and torque were measured using a sensor (Make: Digitech, Model: MP31C09). Figure 1 shows the setup of the machine. A suitable fixture comprising two angle plates was designed and mounted in the machine to hold the rectangular block shaped Inconel 718 workpiece. Machining was conducted in three different conditions, viz., dry, air cooling and nanoemulsion based MQL cooling. The machining variables were spindle speed (N), feed rate (f_r) and depth of cut (d). Milling was conducted at three spindle rotations, viz., 233, 360 and 462 revolutions per minute (RPM). Two feed rates were employed, viz., 13 and 20 mm/min. Depth of cut values used were 0.1 mm and 0.2 mm. For air cooling conditions, an air compressor was utilized to supply air at 0.5, 1.0 and 1.5 bar pressure. During the experiments, compressed air was continuously delivered to the machining zone through a flow regulator equipped with a 3 mm diameter nozzle. The nozzle was placed at a distance of 20 mm from the tip of the cutting tool. In MQL condition, the machining was done using base emulsion (100 ml water + 20 ml Putranjiva Roxburghii oil) and three distinct nanoemulsions based cutting fluids as described earlier. Details of MQL setup are described in Sharma et al (2024).

A statistical analysis was performed to evaluate the distinct influence of each input parameter on force and torque. The investigation included two levels for each of the spindle speed, feed rate, and depth of cut. One is the low level corresponding to minimum value of a parameter and the other is high level corresponding to maximum value of a parameter. The parameter (or factor) that has higher effect on force or torque could be determined. The effect of a factor is evaluated as (Kohli and Dixit, 2005; Dixit and Dixit, 2008) as:

$$\text{Effect of a factor} = \frac{\sum \text{responses at high levels} - \sum \text{responses at low levels}}{\text{half the number of runs in the experiment}} \tag{6.1}$$

6.3 Results and Discussion

This section presents the findings of experimental investigation. Section 3.1 and 3.2 discuss the thermophysical properties and wettability properties of the nanoemulsions, respectively. Next, the results for tool and workpiece temperatures, surface roughness and tool wear during milling of Inconel 718 under nanoemulsion based MQL conditions are presented in Section 3.3, 3.4, and 3.5, respectively. Section 3.6 provides a comparative analysis of axial forces obtained during face milling of Inconel 718 under dry, air-cooled and nanoemulsion based MQL conditions.

6.3.1 Thermophysical properties of nanoemulsions

The effect of SiO_2 and graphene nanoparticles on thermal conductivity was evaluated by analyzing the nanoemulsions. Sharma et al. (2024) reported that the base emulsion comprising 100 ml water and 20 ml Putranjiva roxburghii oil had a thermal conductivity of 0.574 W/m.K. Incorporating 0.5 wt.% SiO_2 nanoparticles into the base fluid enhanced its thermal conductivity to 0.582 W/m·K. indicating an improvement of 1.4%. Further enhancement was observed with the addition of 0.75 wt.% graphene in the base fluid. The thermal conductivity reached 0.622 W/m.K demonstrating an 8.36% enhancement as compared to the base sample. When 0.5 wt. % SiO_2 and 0.5 wt.% graphene were combined, the thermal conductivity reached the highest value of 0.661 W/m.K indicating a 15.16%

enhancement over the base fluid. The combination of SiO_2 and graphene nanoparticles resulted in the highest improvement in thermal conductivity. The percentage relative range (with respect to the mean value) of thermal conductivity measurements was found to be 10.41% for the base oil, 6.87% for the 0.5% SiO_2-based nanoemulsion, 23.79% for the graphene-based nanoemulsion, and 6.65% for the hybrid nanofluid. These values indicate comparatively higher consistency in thermal conductivity for the hybrid and SiO_2-based nanoemulsions, while the graphene-based nanoemulsion exhibited greater variability.

The volumetric heat capacity of the nanoemulsion samples was analysed to evaluate the impact of SiO_2 and graphene nanoparticles. Sharma et al. (2024) reported that the base emulsion comprising 100 ml water and 20 ml Putranjiva roxburghii oil exhibited a volumetric heat capacity of 3.72 $MJ/m^3.K$. Addition of 0.5 wt.% SiO_2 nanoparticles in the base oil increased the volumetric heat capacity to 3.83 $MJ/m^3.K$ indicating an enhancement of 2.96% over the base sample. The volumetric heat capacity reached 3.89 $MJ/m^3.K$, on addition of 0.75 wt.% graphene indicating 4.57% improvement compared to the base emulsion. The highest volumetric heat capacity of 4.12 $MJ/m^3.K$ was obtained when 0.5 wt.% SiO_2 and 0.5 wt.% graphene were added together. The enhancement is 10.75% over the base emulsion. It was observed that graphene contributed more significantly to the increase in volumetric heat capacity than SiO_2. However, their combination provided the most effective enhancement. The percentage relative range of volumetric heat capacity was 12.5% for the base bio-oil, 17.88% for the SiO_2 -based nanoemulsion, 17.76% for the graphene-based nanoemulsion, and 10.99% for the hybrid nanoemulsion. Among all samples, the hybrid nanoemulsion showed the most consistent values, while the SiO_2 and graphene-based nanoemulsions exhibited higher variability.

6.3.2 Wettability study of nanoemulsions

The wettability of nanoemulsions on Inconel 718 was analysed using contact angle measurements as shown in the Figure 6.1. The base emulsion (100 ml water + 20 ml Putranjiva roxburghii oil) exhibited an average contact angle of 46.08°. The addition of 0.5 wt.% SiO_2 nanoparticles increased the contact angle to 57.15°, indicating a increase in wettability. However, addition of 0.75 wt. % graphene nanoparticles reduced the contact angle to 43.36°, demonstrating improved wettability due to graphene's ability to enhance the lubricant's spreading behaviour.

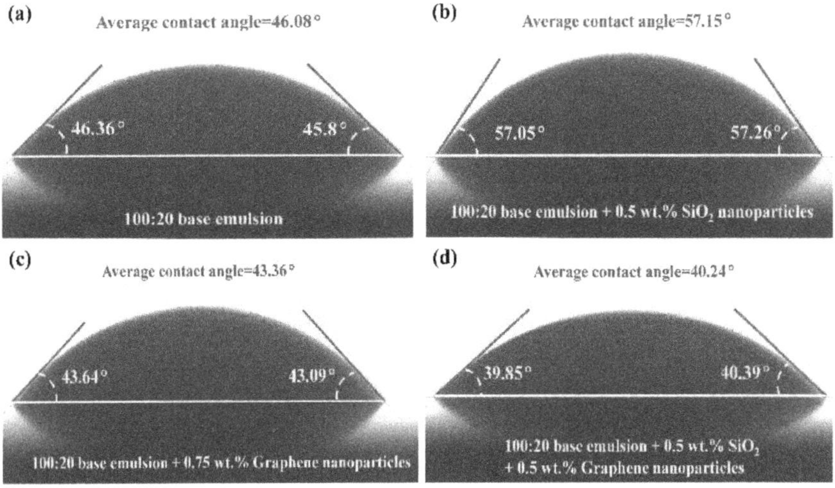

FIGURE 6.1 Contact angle of cutting fluids: (a) base emulsion, (b) SiO_2 based nanoemulsion, (c) Graphene based nanoemulsion, and (d) hybrid (SiO_2 + graphene) nanoemulsion

When 0.5 wt.% SiO_2 and 0.5 wt.% graphene nanoparticles were combined, the contact angle further decreased to 40.24°, indicating an improved spreading of the lubricant on Inconel 718. This combination enhanced the wettability by 12.67% as compared to base emulsion.

6.3.3 Tool and workpiece temperature analysis of nanoemulsions

The variation of tool and workpiece temperature at different cutting conditions are shown in Figure 3. A comparative analysis with dry, air cooling and the base emulsion is carried out. The tool and workpiece temperature for dry machining, air cooling (at 1.5 bar) and base emulsion was reported by (Sharma et al. 2024). When 0.5 wt.% SiO_2 nanoparticles were added in the base emulsion, a slight reduction in temperature was observed, with tool and workpiece temperatures decreasing to 40.5 °C and 30.7 °C, respectively. The inclusion of 0.75 wt.% graphene nanoparticles further reduced the tool temperature to 39.44 °C and the workpiece temperature to 29.9 °C, indicating graphene's superior thermal conductivity and heat dissipation properties. However, when 0.5 wt.% SiO_2 and 0.5 wt.% graphene nanoparticles were added together, the tool and the workpiece temperature increased to 41.66 °C and 32.1 °C, respectively.

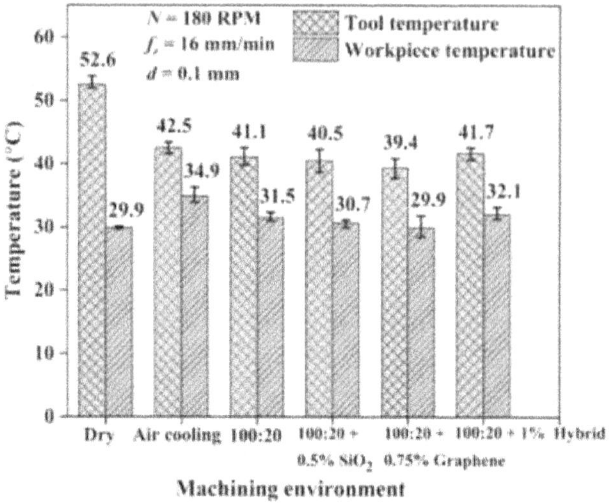

FIGURE 6.2 Variation of tool and workpiece temperature with different machining environments

6.3.4 Study of surface roughness of nanoemulsions

The variation of surface roughness at different cutting environments are shown in Figure 4. Surface roughness at dry machining, air cooling (at 1.5 bar) and base emulsion are 0.64 μm, 0.54 μm and 0.29 μm, respectively as reported by (Sharma et al. 2024). When nanoparticles are added to the base emulsion, further enhancement in surface quality is achieved. With addition of 0.5 wt.% SiO_2 nano-emulsion, the roughness decreased to 0.25 μm. The lowest surface roughness of 0.22 μm was achieved with 0.75% graphene-based nanoemulsion, which can be attributed to graphene's excellent thermal conductivity and superior lubricity. These properties lead to reduced tool wear and enhanced surface integrity. In contrast, the 1% hybrid nano-emulsion shows only a slightly increased surface roughness of 0.30 μm, possibly due to particle agglomeration or instability in dispersion, which might impair lubrication effectiveness and slightly deteriorate the surface finish.

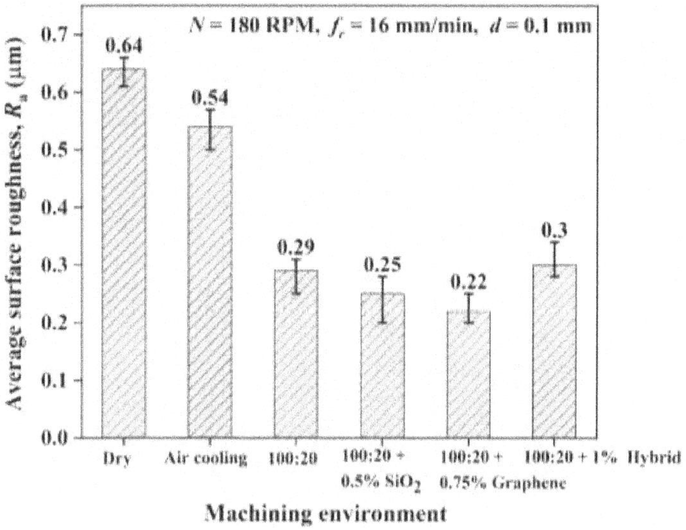

FIGURE 6.3 Variation of average surface roughness with different machining environment

6.3.5 Study of tool wear in nanoemulsions

The wear characteristics of the cutting tool under different MQL conditions using nanoemulsions are presented in Figure 5. In Figure 5(a), formation of built-up edge (BUE) were observed when using a nanoemulsion containing 0.5 wt.% SiO_2 nanoparticles. The lubrication provided by SiO_2 nanoparticles was not sufficient to effectively prevent material adhesion. Figure 5(b) illustrates the wear behaviour with 0.75 wt.% graphene nanoparticles, where flank wear is evident. The superior thermal conductivity and lubricating properties of graphene helped minimize adhesive wear and suppressed BUE formation. In Figure 5(c), the hybrid nanoemulsion comprising 0.5 wt.% SiO_2 and 0.5 wt.% graphene nanoparticles resulted in a combination of adhesive and crater wear.

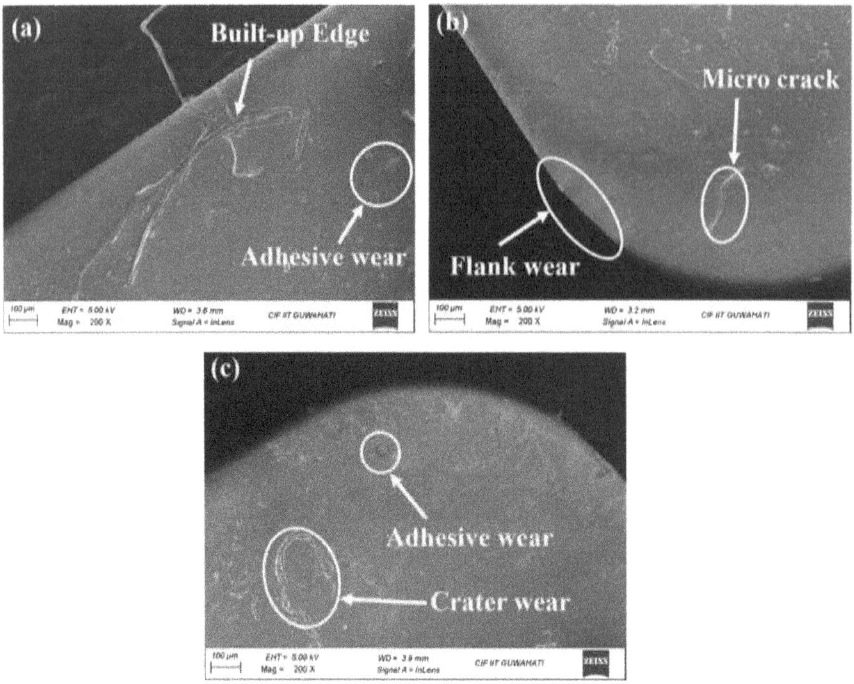

FIGURE 6.4 FESEM images of tool wear at different MQL environments using nanoemlusion: (a) 0.5 wt.% SiO_2, (b) 0.75 wt.% graphene, (c) 0.5 wt.% SiO_2 and 0.5 wt.% graphene

6.4 Force Analysis in Different Cutting Environments

This section presents the analysis of axial forces and torque obtained during face milling of Inconel 718 measured in different machining conditions, viz., dry, air cooling and MQL environments.

6.4.1 Dry machining condition

Figure 6.5 illustrates the time-dependent variation of axial force (F_z) during dry milling under different machining setups. As the cutting tool enters and exits the workpiece, fluctuations in force are observed. Specifically, Figure 6.5 (a) presents the axial force trend at 233 RPM. At this spindle speed, the lowest average value of force, viz., 240.8 N was observed at 13 mm/min feed, 0.1 mm depth of cut. It indicates stable cutting with less resistance. Increasing the value of d_c to 0.2 mm at the same value of feed raised the axial force to 494.2 N. At dry machining across all machining parameters, the lowest value of axial force was 182.9 N; the measured average torque was 39.1 N.mm with power consumption of 1.47 W. This was obtained at *N* 360 RPM, *f* 13 mm/min and *d* 0.1 mm. Complete details of machining parameters and corresponding outputs under dry machining conditions are presented in Table 6.1. The highest axial force recorded under dry conditions was 511.6 N, observed at *N* 233 RPM, *f* 20 mm/min, and d_c 0.2 mm, indicating substantial cutting resistance under this combination of parameters. In another investigation focused on milling of Inconel X-750, Şirin and Kivak (2021) reported the highest cutting force of 498 N in dry conditions.

TABLE 6.1 Machining paramcters and output in dry machining

Sl. No.	Speed, N (RPM)	Feed rate, f (mm/min)	Depth of cut, d_c (mm)	Average axial force, F_z (N)	Average torque, T (N.mm)	Power (W)
1	233	13	0.1	240.8	52.2	1.27
2	233	13	0.2	494.2	49	1.2
3	233	20	0.1	393.2	46.1	1.1
4	233	20	0.2	511.6	51.4	1.25
5	360	13	0.1	344.3	52.6	1.98
6	360	13	0.2	313.2	48.2	1.82
7	360	20	0.1	182.9	39.1	1.47
8	360	20	0.2	504.1	49.7	1.87
9	462	13	0.1	282.8	45.9	2.22
10	462	13	0.2	470.8	40	1.94
11	462	20	0.1	447.9	52.4	2.54
12	462	20	0.2	455.3	43.3	2.1

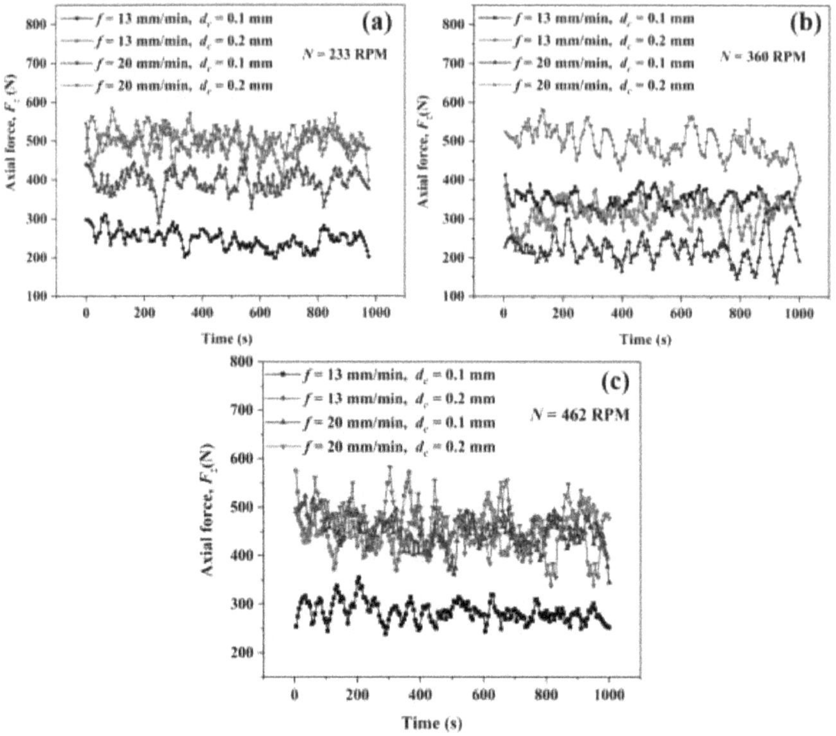

FIGURE 6.5 Variation of axial force with time in dry machining: (a) 233 RPM, (b) 360 RPM, and (c) 462 RPM

For finding the effect of individual factors in dry machining, three factors, viz., N, f and d_c, at two levels are considered. Hence, the total number of combinations is 2^3, i.e., 8. The combinations and effect of factors are shown in Table 6.2. The letters H and L correspond to high and low value of a factor, respectively. It is observed from Table 6.2 that depth of cut has the most significant influence on axial force followed by feed rate. The influence of spindle speed is low. It signifies that increasing the depth of cut substantially raises the cutting resistance. However, the impact of these parameters on torque is relatively minor. Both spindle speed and depth of cut show a negative effect on torque, while feed rate exhibits a small positive effect. These findings suggest that while axial force is highly sensitive to depth of cut and feed rate, but torque remains relatively stable across these parameters.

TABLE 6.2 Effect of factors on force and torque in dry machining

Sl. No.	**Combination**	**Speed, *N* (RPM)**	**Feed rate, *f* (mm/min)**	**Depth of cut, d_c (mm)**	**Average axial force, F_z (N)**	**Average torque, *T* (N.mm)**
1	HHH	462	20	0.2	455.3	43.3
2	HHL	462	20	0.1	447.9	52.4
3	HLH	462	13	0.2	470.8	40
4	LHH	233	20	0.2	511.6	51.4
5	HLL	462	13	0.1	282.8	45.9
6	LHL	233	20	0.1	393.2	46.1
7	LLH	233	13	0.2	494.2	49
8	LLL	233	13	0.1	240.8	52.2
Effect of *N*, *f*, and d_c on F_z, respectively		4.25	79.85	141.8		
Effect of *N*, *f*, and d_c on *T*, respectively		–4.28	1.53	–3.23		

6.4.2 Air cooled machining

The variation in axial force with respect to time in air cooled machining is shown in Figure 6.6. The effect of varying air pressure *P* on axial force, torque and power consumption during face milling of Inconel 718 was investigated under different machining parameters. At a constant speed and feed of 233 RPM and 13 mm/min, respectively, increasing the air pressure *p* led to a moderate rise in axial force. The trend of force variation is shown in Figure 6.6. The average value of F_z increased from 311.4 N (at 0.5 bar) to 353.4 N (at 1.5 bar). Details of machining parameters and their corresponding output in air cooling are presented in Table 6.3. At a feed rate of 20 mm/min, the axial force decreased from 331.9 N at 0.5 bar to 220.9 N at 1 bar. The lowest value of axial force, i.e., 220.9 N was recorded at *N* 233 RPM, *f* 20 mm/min, d_c 0.1 mm and *p* 1 bar; the torque was 25.7 N.mm and power 0.52 W. At 360 RPM, the highest value of average axial force was 551.2 N at *f* 20 mm/min feed, d_c 0.2 mm depth of cut and *p* 1.5 bar. The variations of force are shown in Figures 6.6 (c, d). At 462 RPM, the axial force ranged from 312.2 N to 461.3 N across different combinations of feed rate, depth of cut, and air pressure. The force variation at 462 RPM is illustrated in Figure 6.7. Overall, the effect of air pressure was complex and varied depending on the combination of spindle speed, feed rate, and depth of cut.

TABLE 6.3 Machining parameters and output in air cooling

Sl. No.	Speed, v (RPM)	Feed rate, f (mm/min)	Depth of cut, d_c (mm)	Air pressure, p (bar)	Average axial force, F_z (N)	Average torque, T (N.mm)	Power (W)
1	233	13	0.1	0.5	311.4	45.2	1.10
2	233	13	0.1	1	319.1	48.9	1.19
3	233	13	0.1	1.5	353.4	52	1.27
4	233	13	0.2	0.5	319.8	55	1.34
5	233	13	0.2	1	326.2	56.2	1.37
6	233	13	0.2	1.5	380.4	55	1.34
7	233	20	0.1	0.5	331.9	49	1.20
8	233	20	0.1	1	220.9	25.7	0.63
9	233	20	0.1	1.5	272.8	21.1	0.52
10	233	20	0.2	0.5	355.1	48.5	1.18
11	233	20	0.2	1	323.4	52.1	1.27
12	233	20	0.2	1.5	371.2	56.8	1.39
13	360	13	0.1	0.5	527.3	51.2	1.93
14	360	13	0.1	1	375.4	44	1.66
15	360	13	0.1	1.5	484.7	47.6	1.80
16	360	13	0.2	0.5	423.8	49.6	1.87
17	360	13	0.2	1	485.1	60.6	2.29
18	360	13	0.2	1.5	453.4	17.8	0.67
19	360	20	0.1	0.5	485.1	60.7	2.29
20	360	20	0.1	1	519.6	68.6	2.59
21	360	20	0.1	1.5	478.5	48.2	1.82
22	360	20	0.2	0.5	436.5	43.1	1.63
23	360	20	0.2	1	467.2	51	1.92
24	360	20	0.2	1.5	551.2	56.3	2.12
25	462	13	0.1	0.5	387.9	51.6	2.50
26	462	13	0.1	1	449.8	26.5	1.28
27	462	13	0.1	1.5	370.6	45.6	2.21
28	462	13	0.2	0.5	426.4	55.1	2.67
29	462	13	0.2	1	379.9	19.3	0.93
30	462	13	0.2	1.5	377.7	4.6	0.22
31	462	20	0.1	0.5	312.2	43.9	2.12
32	462	20	0.1	1	408.1	55.6	2.69
33	462	20	0.1	1.5	411.1	61	2.95
34	462	20	0.2	0.5	461.3	43.7	2.12
35	462	20	0.2	1	455.6	37.3	1.81
36	462	20	0.2	1.5	434.3	24.4	1.18

In air cooling assisted machining, four factors, viz., N, f, d and p at two levels are considered. Hence, the total number of combinations is 2^4, i.e., 16. The combinations and effect of factors in air cooling are shown in Table 6.4.

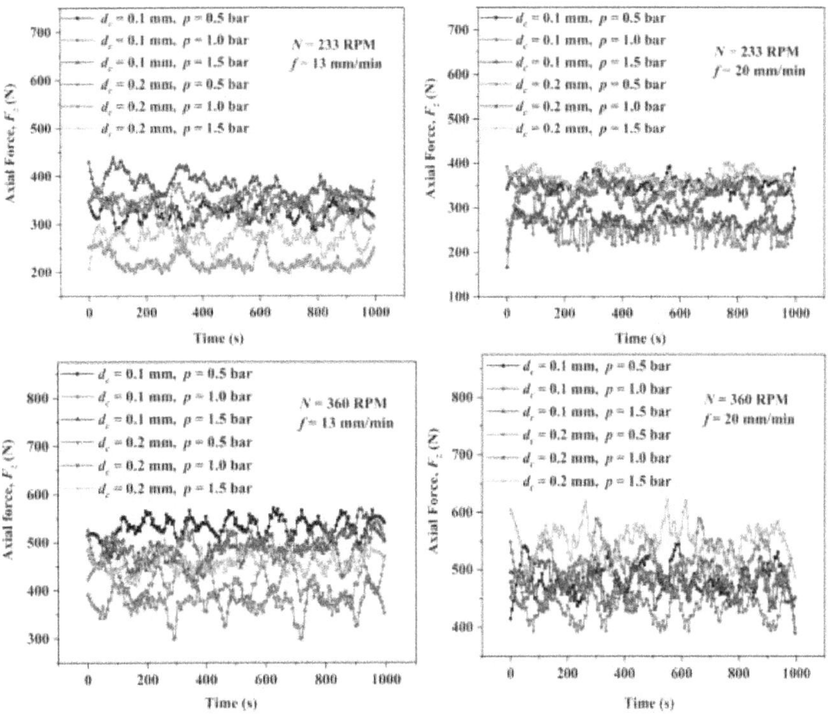

FIGURE 6.6 Variation of axial force with time in air cooling: (a) 233 RPM, 13 mm/min, (b) 233 RPM, 20 mm/min, (c) 360 RPM, 13 mm/min, and (d) 360 RPM, 20 mm/min

It is observed that spindle speed and depth of cut had significant influence on the average axial force. The effect of air pressure on axial force is also noteworthy, with a value of 8.19 N. Increased air pressure enhances cooling and chip evacuation, reducing adhesion and built-up edge (BUE) formation, which in turn reduces cutting resistance. The feed rate shows the least impact on axial force, with an effect of 2.79 N. While feed rate typically increases chip load and thus cutting force, in this case, the influence is marginal likely due to the compensating effects of improved chip breakage and dynamic chip formation under air-cooling.

In contrast, torque responds differently to changes in the input parameters. The feed rate exhibits the most substantial effect on torque, with a value of −5.71 N·mm, followed closely by air pressure (−5.19 N·mm) and spindle speed (−2.84 N·mm). Notably, all three show negative effects, suggesting that increasing these parameters results in reduced torque. This trend can be attributed to changes in chip formation mechanisms and thermal effects.

Higher feed rates reduce the time the tool is engaged per revolution and can cause segmented chip formation, thereby reducing frictional resistance and required torque. Similarly, higher air pressure contributes to better chip evacuation and reduced friction, leading to lower torque values.

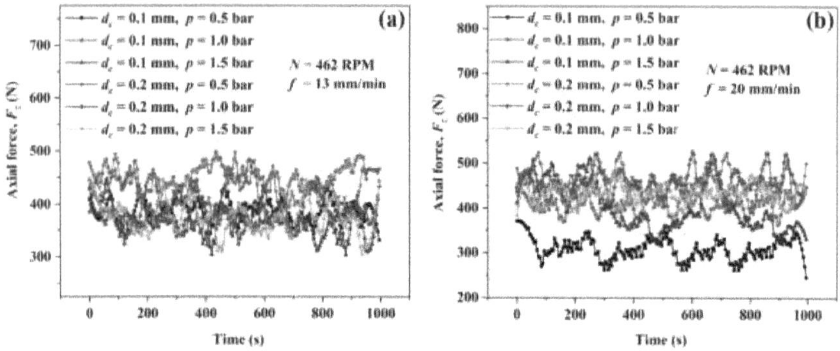

FIGURE 6.7 Variation of axial force with time in air cooling at: (a) 462 RPM, 13 mm/min, (b) 462 RPM, 20 mm/min

This suggests that while axial force is primarily driven by speed and depth of cut. However, torque is less sensitive to individual factors under air-cooled conditions and may be more influenced by their interactions.

TABLE 6.4 Effect of factors on force and torque in air cooling

Sl. No.	**Combination**	**Speed, *N* (RPM)**	**Feed rate, *f* (mm/min)**	**Depth of cut, d_c (mm)**	**Air pressure, *P* (Bar)**	**Average Axial Force, F_z (N)**	**Average Torque, *T* (N.mm)**
1	HHHH	462	20	0.2	1.5	434.3	24.4
2	HHHL	462	20	0.2	0.5	461.3	43.7
3	HHLH	462	20	0.1	1.5	411.1	61
4	HHLL	462	20	0.1	0.5	312.2	43.9
5	HLHH	462	13	0.2	1.5	377.7	34.6
6	HLHL	462	13	0.2	0.5	426.4	55.1
7	HLLH	462	13	0.1	1.5	370.6	45.6
8	HLLL	462	13	0.1	0.5	387.9	51.6
9	LHHH	233	20	0.2	1.5	371.2	56.8

10	LHHL	233	20	0.2	0.5	355.1	48.5
11	LHLH	233	20	0.1	1.5	272.8	21.1
12	LHLL	233	20	0.1	0.5	331.9	49
13	LLHH	233	13	0.2	1.5	380.4	55
14	LLHL	233	13	0.2	0.5	319.8	55
15	LLLH	233	13	0.1	1.5	353.4	52
16	LLLL	233	13	0.1	0.5	311.4	45.2
Effect of N, f, d_c and P on F_z, respectively		60.69	2.79	46.86	8.19		
Effect of N, f, d_c and P on T, respectively		−2.84	−5.71	0.46	−5.19		

6.4.3 Nanoemulsion based cooling with MQL technique

The base emulsion comprising 100 ml water and 20 ml Putranjiva roxburghii oil, acts as a standard cutting fluid without nanoparticles. At a spindle speed of 233 RPM (Figure 6.8), the hybrid (SiO_2 + graphene) and graphene-based nanoemulsions exhibited lower axial forces, particularly at lower depths of cut and feed indicating improved lubrication performance. SiO_2 based nanoemulsions also performed better than the base emulsion. The lowest average axial force at 233 RPM was 177.9 N using hybrid nanoemulsion. The corresponding toque and power consumption is 17.8 N.mm and 0.43 W, respectively. Complete details of machining parameters and their corresponding output in MQL is presented in Table 6.5.

TABLE 6.5 Machining parameters and output in MQL environment

Sl. No.	Speed, N (RPM)	Feed rate, f (mm/min)	Depth of cut, d_c (mm)	Emulsion#	Average axial force, F_z (N)	Average torque, T (N.mm)	Power (W)
1	233	13	0.1	B	545.2	25.4	0.62
2	233	13	0.1	S	389.2	57.6	1.41
3	233	13	0.1	G	315.9	15.2	0.37
4	233	13	0.1	H	177.9	17.8	0.43
5	233	13	0.2	B	393.9	53.4	1.30
6	233	13	0.2	S	333.5	55.7	1.36
7	233	13	0.2	G	279.4	54.7	1.34
8	233	13	0.2	H	281.5	57.5	1.40
9	233	20	0.1	B	431.5	63.2	1.54

10	233	20	0.1	S	362.2	15.7	0.38
11	233	20	0.1	G	287.6	67	1.64
12	233	20	0.1	H	203.6	54.8	1.34
13	233	20	0.2	B	379.4	42.8	1.04
14	233	20	0.2	S	389.2	48.2	1.18
15	233	20	0.2	G	314.6	63.6	1.55
16	233	20	0.2	H	320.9	56.3	1.37
17	360	13	0.1	B	432.4	51.3	1.93
18	360	13	0.1	S	390.3	54.9	2.07
19	360	13	0.1	G	242.1	22.5	0.85
20	360	13	0.1	H	274.7	54.7	2.06
21	360	13	0.2	B	392.8	50.2	1.89
22	360	13	0.2	S	390.3	54.2	2.04
23	360	13	0.2	G	281	61.2	2.31
24	360	13	0.2	H	528.7	43.4	1.64
25	360	20	0.1	B	464.5	60.5	2.28
26	360	20	0.1	S	358.3	54.6	2.06
27	360	20	0.1	G	331.7	57.3	2.16
28	360	20	0.1	H	324	60.5	2.28
29	360	20	0.2	B	375.1	44.1	1.66
30	360	20	0.2	S	419.6	56.8	2.14
31	360	20	0.2	G	292.1	57.7	2.18
32	360	20	0.2	H	586.9	44.1	1.66
33	462	13	0.1	B	408.7	57.5	2.78
34	462	13	0.1	S	353.4	44.1	2.13
35	462	13	0.1	G	328.1	56.5	2.73
36	462	13	0.1	H	346.2	51.8	2.51
37	462	13	0.2	B	365.2	21.43	1.04
38	462	13	0.2	S	263.7	45.6	2.21
39	462	13	0.2	G	290.4	52.2	2.53
40	462	13	0.2	H	286.3	59.4	2.87
41	462	20	0.1	B	446.1	51.7	2.50
42	462	20	0.1	S	209.3	44.9	2.17
43	462	20	0.1	G	219.5	17.7	0.86
44	462	20	0.1	H	257.2	52.5	2.54
45	462	20	0.2	B	386.6	60	2.90
46	462	20	0.2	S	279	18.1	0.88
47	462	20	0.2	G	249.1	19.67	0.95
48	462	20	0.2	H	291.4	22.2	1.07

#**B:** base emulsion, **S:** SiO_2 based nanoemulsion, **G:** Graphene based nanoemulsion, **H:** hybrid nanoemulsion

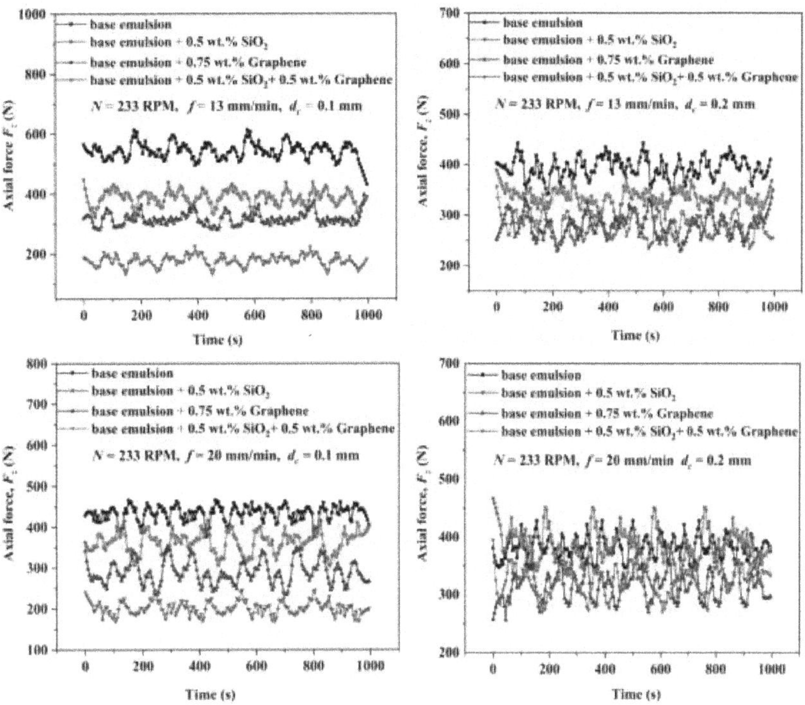

FIGURE 6.8 Variations of axial milling force with time in MQL environments at 233 RPM: (a) 13 mm/min, 0.1 mm, (b) 13 mm/min, 0.2 mm, (c) 20 mm/min, 0.1 mm, and (d) 20 mm/min, 0.2 mm

Figures 6.9 and 6.10 show the axial force variation at different machining parameters in MQL based condition. At both 360 and 462 RPM spindle speeds, the performance of different MQL emulsions significantly influenced the axial force and torque during face milling. At 360 RPM, the hybrid and graphene based nanoemulsions generally showed lower axial forces and torque compared to the base and SiO_2 based nanoemulsions. SiO_2 based nanoemulsion resulted in the lowest force, viz., 209.3 N at *N* 462 RPM, *f* 20 mm/min and *d* 0.1 mm. This corresponds to average torque and power of 44.9 N.mm and 2.17 W, respectively. Overall, nanoemulsions demonstrated superior lubrication characteristics at elevated spindle speeds, contributing to lower axial forces and enhanced machining performance.

In MQL based milling, three factors, viz., *N*, *f*, and d_c at two levels are considered for each MQL conditions. The combinations and effect of factors in base oil are shown in Table 6.6. Under MQL using a base emulsio, depth of cut had the greatest impact on reducing axial force (−76.6 N), due to improved lubrication and reduced friction. Spindle speed also lowered axial force (−35.85 N) through thermal softening of Inconel 718, while feed rate had minimal effect (−17.35 N) due to better chip flow.

For torque, feed rate had the most significant positive effect (+14.99 N·mm) as higher material removal increased load. Spindle speed had a minor positive influence (+1.46 N·mm), while depth of cut slightly reduced torque (−5.04 N·mm). Overall, the emulsion-based MQL enhanced machining by lowering both axial force and torque through consistent lubrication.

TABLE 6.6 Effect of factors on force and torque in base emulsion environment

Base emulsion						
Sl. No.	**Combination**	**Speed, *N* (RPM)**	**Feed rate, *f* (mm/min)**	**Depth of cut, d_c (mm)**	**Average axial force, F_z (N)**	**Average torque, *T* (N.mm)**
1	HHH	462	20	0.2	386.6	60
2	HHL	462	20	0.1	446.1	51.7
3	HLH	462	13	0.2	365.2	21.43
4	LHH	233	20	0.2	379.4	42.8
5	HLL	462	13	0.1	408.7	57.5
6	LHL	233	20	0.1	431.5	63.2
7	LLH	233	13	0.2	393.9	53.4
8	LLL	233	13	0.1	545.2	25.4
Effect of *N*, *f*, and d_c on F_z, respectively		-35.85	-17.35	-76.6		
Effect of *N*, *f*, and d_c on *T*, respectively		1.46	14.99	-5.04		

Using MQL with a 0.5 wt.% SiO_2 nanoparticle-based emulsion during face milling of Inconel 718 led to a notable decrease in axial force and a moderate effect on torque (Table 6.7). Spindle speed had the strongest impact on reducing axial force (−92.18 N), attributed to thermal softening and the lubricating film formed by the nanoparticles, which reduced friction and adhesion.

Feed rate also reduced axial force (−25.03 N), likely due to improved chip flow and heat dissipation from the nano-enhanced fluid. Depth of cut had the least effect (−12.18 N), indicating stable lubrication even under heavier loads. For torque, feed rate showed the highest increase (+19.03 N·mm), as expected from greater chip loads. However, the nanoparticle emulsion helped control torque rise through better lubrication and thermal management. Spindle speed slightly reduced torque (−6.13 N·mm), and depth of cut showed a minor increase (+1.33 N·mm), both suggesting controlled tool–work interaction under nano-lubrication.

TABLE 6.7 Effect of factors on force and torque in SiO_2 based nanoemulsion

Base emulsion + 0.5 wt.% SiO_2 nanoparticles						
Sl. No.	**Combination**	**Speed, N (RPM)**	**Feed rate, f (mm/min)**	**Depth of cut, d_c (mm)**	**Average axial force, F_z (N)**	**Average torque, T (N.mm)**
1	HHH	462	20	0.2	279	18.1
2	HHL	462	20	0.1	209.3	44.9
3	HLH	462	13	0.2	263.7	45.6
4	LHH	233	20	0.2	389.2	48.2
5	HLL	462	13	0.1	353.4	44.1
6	LHL	233	20	0.1	362.2	15.7
7	LLH	233	13	0.2	333.5	55.7
8	LLL	233	13	0.1	389.2	57.6
Effect of N, f, and d_c on F_z, respectively		–92.18	–25.03	–12.18		
Effect of N, f, and d_c on T, respectively		–6.13	–19.03	1.33		

Effect of factors of 0.75 wt.% graphene-based nanoemulsion significantly reduced both axial force and torque shown in Table 6.8. Feed rate had the greatest effect on lowering axial force (−35.75 N), owing to graphene's superior lubrication and load-bearing capacity. Spindle speed also reduced force (−27.6 N) due to thermal softening and graphene's high thermal conductivity, while depth of cut had a minimal impact (−4.4 N), indicating consistent lubrication. For torque, depth of cut slightly increased resistance (+8.44 N·mm), whereas spindle speed and feed rate lowered torque (−13.61 and −2.66 N·mm). Overall, the graphene-enriched emulsion improved lubrication stability and heat dissipation, demonstrating its effectiveness in machining hard-to-cut materials.

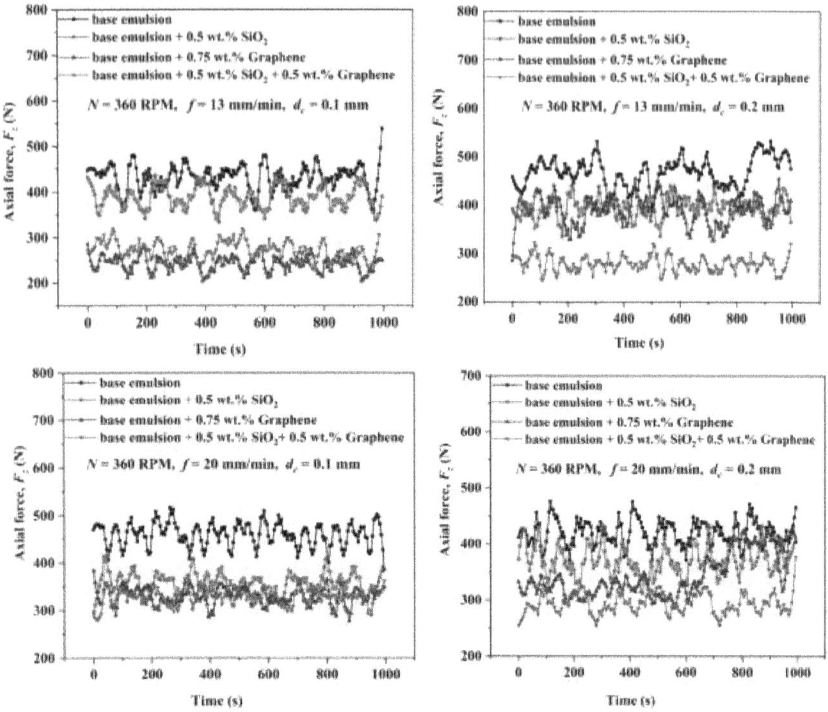

FIGURE 6.9 Variations of axial milling force with time in MQL environments at 360 RPM: (a) 13 mm/min, 0.1 mm, (b) 13 mm/min, 0.2 mm, (c) 20 mm/min, 0.1 mm, and (d) 20 mm/min, 0.2 mm

In hynrid nanoemulsion (0.5 wt.% SiO_2, and 0.5 wt.% graphene nanoparticles), distinct patterns in axial force and torque were observed. Spindle speed (+49.3 N) and depth of cut (+48.8 N) significantly increased axial force, whereas feed rate had a slight negative impact (−4.7 N), likely due to enhanced lubrication and smoother chip flow at higher feeds. The increase in force with speed and depth is linked to greater tool–material interaction and thermal load, partially mitigated by the improved film strength and lubrication from the combined nanoparticles.

In terms of torque, depth of cut had the highest positive effect (+3.38 N·mm), followed by spindle speed (+1.13 N·mm), while feed rate marginally decreased torque (−1.43 N·mm). The synergistic effect of SiO_2 and graphene contributed to stable lubrication, reduced friction, and consistent cutting performance, enhancing overall machining efficiency under MQL conditions.

TABLE 6.8 Effect of factors on force and torque in graphene based nanoemulsion

Base emulsion + 0.75 wt.% graphene nanoparticles						
Sl. No.	**Combination**	**Speed, *N* (RPM)**	**Feed rate, *f* (mm/min)**	**Depth of cut, d_c (mm)**	**Average axial force, F_z (N)**	**Average torque, *T* (N.mm)**
1	HHH	462	20	0.2	249.1	19.67
2	HHL	462	20	0.1	219.5	17.7
3	HLH	462	13	0.2	290.4	52.2
4	LHH	233	20	0.2	314.6	63.6
5	HLL	462	13	0.1	328.1	56.5
6	LHL	233	20	0.1	287.6	67
7	LLH	233	13	0.2	279.4	54.7
8	LLL	233	13	0.1	315.9	15.2
Effect of *N*, *f*, and d_c on F_z, respectively		−27.6	−35.75	−4.4		
Effect of *N*, *f*, and d_c on *T*, respectively		−13.61	−2.66	8.44		

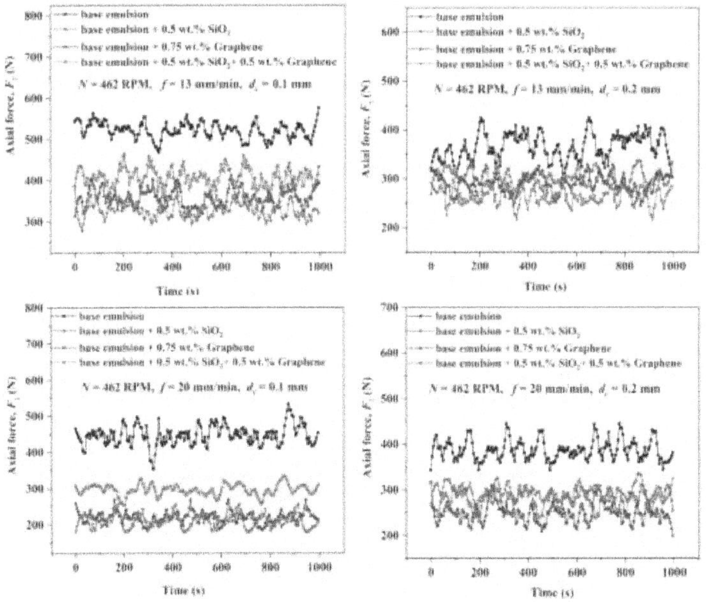

FIGURE 6.10 Variations of axial milling force with time in MQL environments at 462 RPM: (a) 13 mm/min, 0.1 mm, (b) 13 mm/min, 0.2 mm, (c) 20 mm/min, 0.1 mm, and (d) 20 mm/min, 0.2 mm

TABLE 6.9 Effect of factors on force and torque based hybrid nanoemulsion

Base emulsion + 0.5 wt.% SiO_2 nanoparticles + 0.5 wt.% graphene nanoparticles						
Sl. No.	**Combination**	**Speed, *N* (RPM)**	**Feed rate, *f* (mm/min)**	**Depth of cut, d_c (mm)**	**Average axial force, F_z (N)**	**Average torque, *T* (N.mm)**
1	HHH	462	20	0.2	291.4	22.2
2	HHL	462	20	0.1	257.2	52.5
3	HLH	462	13	0.2	286.3	59.4
4	LHH	233	20	0.2	320.9	51.3
5	HLL	462	13	0.1	346.2	51.8
6	LHL	233	20	0.1	203.6	54.8
7	LLH	233	13	0.2	281.5	57.5
8	LLL	233	13	0.1	177.9	17.8
Effect of N, f, and d_c on F_z, respectively		49.3	–4.7	48.8		
Effect of N, f, and d_c on T, respectively		1.13	–1.43	3.38		

As cutting temperature rises, the workpiece softens, which may slightly reduce cutting forces, but higher temperatures also accelerate tool wear, increase built-up edge formation, and destabilize the cutting process, leading to poorer surface finish. Thus, excessive heat ultimately increases surface roughness despite any temporary reduction in cutting forces. It clarifies that an appropriate balance between cutting forces and surface roughness is necessary for optimal machining. Extremely low forces accompanied by poor surface quality, or vice versa, are not acceptable. Effective machining requires controlled temperature, limited tool wear, stable chip formation, and consistent surface integrity.

The hybrid nanoemulsion combines the advantages of both nanoparticles used, resulting in enhanced lubrication and superior thermal conductivity compared to single nanoparticle emulsions. The synergistic interaction between SiO_2 and graphene improves the formation of a stable, low-friction tribofilm at the tool–workpiece interface. This reduces adhesion and minimizes direct metal-to-metal contact, thereby lowering frictional resistance during cutting. Additionally, the higher heat dissipation capability of the hybrid formulation reduces cutting-zone temperature, which in turn decreases material softening, cutting-edge wear, and ploughing forces. These combined effects contribute to smoother chip flow and significantly reduced cutting forces.

6.5 Summary

This study comprehensively evaluated the thermal, tribological, and machining performance of nano-doped bio-based cutting fluids during the milling of Inconel 718. The incorporation of SiO_2 and graphene nanoparticles into Putranjiva roxburghii oil-based emulsions demonstrated significant improvements in thermal conductivity, specific heat, wettability, axial force, tool temperature and surface finish. Among the tested nanoemulsions, the graphene-based cutting fluid showed superior thermal and tribological properties due to its excellent lubricity and heat dissipation capabilities. Meanwhile, the hybrid nanoemulsion (0.5% SiO_2 + 0.5% graphene) offered the best reduction in cutting forces, attributed to the synergistic effects of wear resistance from SiO_2 and lubrication from graphene. Compared to dry and air-cooled conditions, nano-doped bio-emulsions significantly reduced tool and workpiece temperatures, minimized surface roughness, and stabilized axial forces. This enhanced tool life and machining stability. However, results also indicated that improper dispersion or nanoparticle agglomeration, particularly in hybrid emulsions, can negatively affect surface integrity and thermal performance.

References

Bartolomeis, A., Newman, S.T., Jawahir, I.S., Biermann, D. and Shokrani, A., 2021. Future research directions in the machining of Inconel 718. *Journal of Materials Processing Technology*, *297*, p.117260.

Dixit, P.M. and Dixit, U.S., 2008. *Modeling of Metal Forming and Machining Processes: by Finite Element and Soft Computing Methods*. Springer Science & Business Media, London.

Dixit, U.S., Sarma, D.K. and Davim, J.P., 2012. *Environmentally friendly machining*. Springer Science & Business Media.

Hegab, H., Umer, U., Soliman, M. and Kishawy, H.A., 2018. Effects of nano-cutting fluids on tool performance and chip morphology during machining Inconel 718. *The International Journal of Advanced Manufacturing Technology*, *96*, pp.3449–3458.

Kohli, A. and Dixit, U.S., 2005. A neural-network-based methodology for the prediction of surface roughness in a turning process. *The International Journal of Advanced Manufacturing Technology*, *25*, pp.118–129.

Mahesh, K., Philip, J.T., Joshi, S.N. and Kuriachen, B., 2021. Machinability of Inconel 718: A critical review on the impact of cutting temperatures. *Materials and Manufacturing Processes*, *36*(7), pp.753-791.

Mishra, S.K., Gómez Escudero, G., González-Barrio, H., Calleja-Ochoa, A., Pereira Neto, O. and Norberto López de Lacalle Marcaide, L., 2024. Wear mechanisms and cutting performance of coated carbides in milling LPBF IN718 alloy under hybrid cryogenic conditions. *Machining Science and Technology*, *28*(2), pp.187-214.

Rakesh, M. and Datta, S., 2019. Effects of cutting speed on chip characteristics and tool wear mechanisms during dry machining of Inconel 718 using uncoated WC tool. Arabian Journal for Science and Engineering, 44, pp.7423-7440.

Roushan, A., Biswas, S. and Chetan, 2024. Emerging sustainable techniques in metal cutting to reduce the application of metalworking fluids: A review. *Proceedings of the Institution of Mechanical Engineers, Part E: Journal of Process Mechanical Engineering*, p.09544089241256591.

Sharma, D.K., Singh, P.K., Sharma, F. and Dixit, U.S., 2024. Machining performance of Inconel 718 in dry, air-cooling and bio-oil based minimum quantity lubrication environments. *Proceedings of the Institution of Mechanical Engineers, Part E: Journal of Process Mechanical Engineering*, p.09544089241301596.

Sharma, D.K., Singh, P.K., Sharma, F. and Dixit, U.S., 2025. Development and characterization of Putranjiva roxburghii with addition of SiO_2 and graphene nanoparticles for using it as a cutting fluid. *Proceedings of the Institution of Mechanical Engineers, Part C: Journal of Mechanical Engineering Science*, 239(11), pp. 4162-4181.

Shokrani, A., Betts, J. and Jawahir, I.S., 2022. Improved performance and surface integrity in finish machining of Inconel 718 with electrically charged tungsten disulphide MQL. *CIRP Annals*, *71*(1), pp.109-112.

Şirin, Ş. and Kıvak, T., 2021. Effects of hybrid nanofluids on machining performance in MQL-milling of Inconel X-750 superalloy. *Journal of Manufacturing Processes*, *70*, pp.163-176.

Supriya, B., Keerthana, V., Nambirajan, S. and Dharmendira, K.M., 2017. Medicinal values of Putranjiva roxburghii-a review. *International Journal of Current Pharmaceutical Research*, *9*(5), pp.5-8.

Thakur, D.G., Ramamoorthy, B. and Vijayaraghavan, L., 2009. Study on the machinability characteristics of superalloy Inconel 718 during high speed turning. Materials & Design, 30(5), pp.1718-1725.

United Nations. https://sdgs.un.org/goals (Accessed on May 15, 2025)

CHAPTER

7

Conclusions and future scope

7.1 Conclusions

The present research addressed the pressing need for sustainable and high-performance machining solutions by developing and evaluating novel cutting fluids based on *Putranjiva roxburghii* vegetable oil and its nanofluid derivatives for the milling of Inconel 718 a challenging-to-machine, nickel-based superalloy. In this context, the study systematically investigated the development, characterization, and machining application of a novel bio-based cutting fluid derived from *Putranjiva roxburghii* seeds, including its nanofluid formulations, for the milling of Inconel 718 alloy. Inconel 718 is a nickel-based superalloy renowned for its high strength, corrosion resistance, and performance under elevated temperatures, which makes it indispensable in aerospace, marine, and nuclear industries.

Putranjiva roxburghii oil, extracted using a cold-press method to retain its natural properties, was chosen as the base cutting fluid due to its renewable origin and bio-compatibility. The fluid was then augmented with nanoparticles of silicon dioxide (SiO_2) and graphene in varying concentrations (0.25%, 0.5%, 0.75%, and 1% by weight), both individually and in hybrid (equal proportion) forms, to improve its heat dissipation, lubrication, and wetting properties. The results from the characterization of nanofluids concluded that:

- ❖ Addition of SiO_2 and graphene nanoparticles enhanced the thermal conductivity of the nanofluid. The highest thermal conductivity was observed at 0.75 wt.% of graphene-dispersed nanofluids.

- ❖ Nanofluids caused a slight increase in viscosity, which contributed to improved lubricity. Higher viscosity supported better lubrication during machining.

- ❖ Wettability studies on Inconel 718 revealed that the addition of nanoparticles lowered the contact angle. The hybrid nanofluid containing both graphene and SiO_2 showed the most desirable wettability characteristics. No single nanoparticle concentration was found to be ideal for all performance aspects.

- ❖ The incorporation of nanoparticles into the base oil significantly enhanced machining performance. Specifically, the use of hybrid nanofluid at a spindle speed of 180 rpm resulted in a 23% reduction in surface roughness compared to dry machining

- ❖ Tool temperature dropped from 57.9°C (dry) to 49.8°C with graphene-based nanofluid. These results demonstrate the superior performance of nanoparticle-enhanced fluids in improving overall machining outcomes.

In the machining study, the milling behaviour of Inconel 718 was investigated using a cemented carbide tool under three different cooling conditions: dry machining, air-cooling, and bio-based minimum quantity lubrication (MQL). The results from the milling performance in different cutting environments concluded that:

- ❖ A detailed investigation on dry milling of Inconel 718 revealed that machining performance is highly sensitive to process parameters. Optimization using Box's evolutionary algorithm was employed to determine the best combination of input variables.
- ❖ The minimum surface roughness value recorded was 0.29 μm under MQL conditions. This represents a 54.7% reduction compared to dry machining

and a 46.3% reduction compared to air-cooled machining. The significant enhancement in surface finish underscores the superior performance of Putranjiva roxburghii oil as a bio-based cutting fluid.

- Microstructural analysis revealed that dry machining led to abrasive flank wear and substantial formation of built-up edge (BUE). MQL significantly reduced tool wear, indicating improved tool life and better lubrication. For instance, higher air pressure in air cooling helped reduce abrasive wear but led to an increase in crater wear.

Further study systematically evaluated the thermal, tribological, and machining performance of nano-doped bio-based cutting fluids during the milling of Inconel 718. The incorporation of SiO_2 and graphene nanoparticles into *Putranjiva roxburghii* oil-based emulsions led to significant improvements in key performance metrics.

- Among all tested formulations, the graphene-based nanoemulsion exhibited superior thermal and tribological behavior due to its excellent lubricity and heat dissipation characteristics. Additionally, the hybrid formulation containing 0.5 wt.% SiO_2 and 0.5 wt.% graphene provided the greatest reduction in cutting forces, attributed to the combined effects of SiO_2-induced wear resistance and graphene-enhanced lubrication.
- Compared to conventional dry and air-cooled machining, the nano-doped bio-emulsions resulted in substantial reductions in tool and workpiece temperatures, improved surface integrity, and more stable cutting forces, thereby enhancing tool life and machining reliability.

7.2 Future Scope

While the study has successfully demonstrated the potential of bio-based nanofluids in improving machining performance, several avenues remain open for further exploration and industrial implementation:

7.2.1 Application to Other Materials and Processes:

- The efficacy of *Putranjiva roxburghii*-based nanofluids can be extended to other hard-to-machine alloys such as titanium alloys, stainless steels, and ceramics.

- Future research may involve testing under different machining operations such as turning, drilling, and grinding to generalize the applicability of the developed cutting fluids.

7.2.2 Advanced Nanoparticles and Hybrid Structures:

- Incorporation of other high-performance nanoparticles such as carbon nanotubes (CNTs), boron nitride, or metal oxides could be explored for further improving the lubrication and thermal properties.
- Functionalized or surface-treated nanoparticles may enhance dispersion stability and compatibility with bio-oils.

7.2.3 Tool Coating and Geometry Optimization:

- The interaction between tool coatings (e.g., TiAlN, DLC) and bio-nanofluids could be studied to determine synergistic effects on tool life and wear mechanisms.
- Optimization of tool geometry for MQL-based cutting environments could lead to improved chip evacuation, thermal management, and wear resistance.

7.2.4 Life Cycle Assessment (LCA) and Economic Analysis:

- A comprehensive life cycle assessment of *Putranjiva roxburghii* bio-oil production and disposal will help validate its environmental impact compared to mineral oil-based fluids.
- Cost-benefit analysis, considering nanoparticle synthesis, oil extraction, and scalability, will be crucial for commercial adoption.

7.2.5 Real-Time Monitoring and Smart Lubrication Systems:

- Integration of real-time sensing and control systems for adaptive lubrication (smart MQL) can help dynamically adjust fluid supply based on cutting conditions and sensor feedback.
- Application of artificial intelligence and machine learning techniques for predictive modeling of tool wear, surface finish, and fluid consumption can enhance process control.

7.2.6 Tribological and Thermal Modelling:

- Future work may involve numerical simulation and modeling of the heat generation, distribution, and tribological behavior at the tool–chip interface under MQL conditions using finite element or computational fluid dynamics (CFD) tools.
- These models can be validated against experimental results to gain deeper insight into the underlying mechanisms and optimize fluid design.